Studies in Systems, Decision and Control

Volume 424

The series "Studies in Systems, Decision and Control" (SSDC) covers both new developments and advances, as well as the state of the art, in the various areas of broadly perceived systems, decision making and control–quickly, up to date and with a high quality. The intent is to cover the theory, applications, and perspectives on the state of the art and future developments relevant to systems, decision making, control, complex processes and related areas, as embedded in the fields of engineering, computer science, physics, economics, social and life sciences, as well as the paradigms and methodologies behind them. The series contains monographs, textbooks, lecture notes and edited volumes in systems, decision making and control spanning the areas of Cyber-Physical Systems, Autonomous Systems, Sensor Networks, Control Systems, Energy Systems, Automotive Systems, Biological Systems, Vehicular Networking and Connected Vehicles, Aerospace Systems, Automation, Manufacturing, Smart Grids, Nonlinear Systems, Power Systems, Robotics, Social Systems, Economic Systems and other. Of particular value to both the contributors and the readership are the short publication timeframe and the world-wide distribution and exposure which enable both a wide and rapid dissemination of research output.

Indexed by SCOPUS, DBLP, WTI Frankfurt eG, zbMATH, SCImago.

All books published in the series are submitted for consideration in Web of Science.

More information about this series at https://link.springer.com/bookseries/13304

Kumaresan Perumal · Chiranji Lal Chowdhary ·
Logan Chella
Editors

Innovative Supply Chain Management via Digitalization and Artificial Intelligence

Editors
Kumaresan Perumal
School of Information Technology
and Engineering
Vellore Institute of Technology
Vellore, Tamil Nadu, India

Chiranji Lal Chowdhary
School of Information Technology
and Engineering
Vellore Institute of Technology
Vellore, Tamil Nadu, India

Logan Chella
Cyber Innovations LLC
Virginia, VA, USA

ISSN 2198-4182 ISSN 2198-4190 (electronic)
Studies in Systems, Decision and Control
ISBN 978-981-19-0242-0 ISBN 978-981-19-0240-6 (eBook)
https://doi.org/10.1007/978-981-19-0240-6

This Springer imprint is published by the registered company Springer Nature Singapore Pte Ltd.
The registered company address is: 152 Beach Road, #21-01/04 Gateway East, Singapore 189721, Singapore

Contents

About the Editors

Kumaresan Perumal is a Senior Assistant Professor in the School of Information Technology and Engineering (SITE) at Vellore Institute of Technology (VIT) Vellore, India. He received his Bachelor's (ECE) and Ph.D. degree (CSE) from VIT and master's degree (CSE) from Anna University, Chennai. He has extensive expertise in academia and research. His research areas include IoT, WBAN, cloud computing, machine learning, deep learning, supply chain management, and robotic process automation. He has published several research papers in international journals and conferences. He has got his innovation patented on IoT-based Signature Robot. He is a life member of the CSI.

Chiranji Lal Chowdhary is an Associate Professor in the School of Information Technology & Engineering at VIT University, where he has been since 2010. He received a B.E. (CSE) from MBM Engineering College at Jodhpur in 2001, and M. Tech. (CSE) from the M.S. Ramaiah Institute of Technology at Bangalore in 2008. He received his Ph.D. from the VIT University Vellore in 2017. From 2006 to 2010 he worked at M.S. Ramaiah Institute of Technology in Bangalore, eventually as a Lecturer. His research interests span both computer vision and image processing. Much of his work has been on images, mainly through the application of image processing, computer vision, pattern recognition, machine learning, biometric systems, deep learning, soft computing, and computational intelligence. He has given a few invited talks on medical image processing. Professor Chowdhary is editor/co-editor of more than 7 books and is the author of over fifty articles on computer science. He published two patents deriving from his research.

Logan Chella is Chief Executive Officer of Cyber Innovations LLC and an entrepreneur in the US Technology practice. A technology leader and visionary with over 25 years of experience in technology innovation and implementation, he is focused on helping clients transform the way they deliver their missions/businesses through technology. He is responsible for producing cross-industry tech trends research on emerging and disruptive technology that impacts organizations in the near term. He also oversees the portfolio of emerging offerings within technology incubation function including areas like artificial intelligence, quantum computing, IoT, and more. Logan is a trusted advisor to senior US federal executives on the latest technologies enabling them to take advantage of the benefits of emerging technologies. Logan holds a graduate degree from Chennai University and has several master certifications in the USA including Harvard Business School. Logan is also a renowned thought leader in the Federal Information Technology industry and has published studies and white papers. Before his career with cyber innovations, Logan worked for Deloitte Consulting providing IT management and consulting services, and for Society for Worldwide Interbank Financial Telecommunication (SWIFT) as Senior Architect where he led various innovative solutions to transform the global banking industry to the 21st-century services. Logan is a strong proponent of green energy; he commissioned India's first solar power plant as part of the National Solar Mission in 2010.

Digital Supply Chain Management Using AI, ML and Blockchain

Anil Kumar Gupta, Gaurang Vivek Awatade, Suyog Sanjay Padole, and Yash Santosh Choudhari

Abstract Supply Chain Management (SCM) is a kind of network in which each phase of product development is tracked right from gathering raw materials till the delivery of the product. Developing a product, distributing it to suppliers, delivering to customers, collecting feedback, etc., all are tracked and maintained using supply chain management. A secured supply chain should be held to ensure that fake products do not enter into the market. For example; if counterfeit drugs enter the market, then it will have various side effects on patients as well as it will affect the pharmaceutical companies. Hence, a secure supply chain is a must. The significance of SCM has increased as it plays a vital role in the decision-making process. Hence, a proper supply chain should be maintained along with a technology which will also help in the decision-making process. One such technology is Artificial Intelligence (AI). AI is the technology in which machines are programmed to think like humans and perform an action accordingly. Machine Learning can be used in detecting fake products that enter the market. Also, there are various algorithms in AI which can be used in the decision-making process. As the digitalization of SCM is significant, similarly, its security is also worthwhile. SCM can be implemented as a Drug Supply chain Management system, which Blockchain can be integrated to counter the supply of counterfeit drugs into the market by tracking. This chapter discusses Digitalized and AI-based Drug Supply Chain Management. With Supply chain analytics tracking each step of the inventory process: from the loading dock to the supply cabinet, to the patient's hospital room, Hospitals can expect massive Supply chain analytics are used to measure, monitor, and improve individual business processes as well as the overall performance and health of the supply chain. Drug supply chain analytics will enable demand visibility, inventory visibility and Freight analytics.

Keywords Blockchain · Digitalization · Artificial intelligence · Supply chain management · Security

A. K. Gupta (✉)
Centre for Development of Advanced Computing (C-DAC), Pune, India

G. V. Awatade · S. S. Padole · Y. S. Choudhari
Department of Computer, Dr. D.Y. Patil Institute of Technology, SPPU, Pimpri, Pune, India

K. Perumal et al. (eds.), *Innovative Supply Chain Management via Digitalization and Artificial Intelligence*, Studies in Systems, Decision and Control 424,
https://doi.org/10.1007/978-981-19-0240-6_1

1 Introduction

Supply Chain Management (SCM) is a network in which each phase of product development is tracked right from gathering raw materials till the delivery of the product. It is a process that involves planning, controlling and implementing the operations of the supply chain to satisfy customer requirements effectively and efficiently. The subareas of a supply chain comprise of Forecasting, Procurement, Logistics, Operations, Inventory Management, Transport, Warehousing, Distribution, Customer Service etc. Developing a product, distributing it to suppliers, delivering to customers, collecting feedback, etc., all are tracked and maintained using supply chain management. One should maintain a secured supply chain to ensure that fake products do not enter the market. For example; if counterfeit drugs enter the market, then it will have various side effects on patients as well as it will affect the pharmaceutical companies. Hence, a secure supply chain is a must.

The significance of SCM has increased as it plays an essential role in the decision-making process. Therefore, one should maintain a proper supply chain along with a technology which will also help in the decision-making process. One such technology is Artificial Intelligence (AI). AI is the technology in which we program the machines to think like humans and perform an action accordingly. Another branch of AI is Machine Learning (ML) which we can use in detecting fake products that enter the market. Also, there are various algorithms in AI which we can use in the decision-making process. As the digitalization of SCM is significant, similarly, its security is also worthwhile. Blockchain is a revolutionary technology which is used in safety and can be efficiently implemented in SCM. Blockchain is no longer confined to transfer of funds between two accounts as it crawls its way into diverse fields such as IoT, Healthcare, Real Estate and also SCM. In Healthcare, SCM can be implemented as a Drug Supply chain Management system using blockchain, which could counter the supply of fake drugs into the market by tracking.

1.1 Chapter Overview

This chapter contains the introduction of Supply Chain Management and discusses each phase of SCM. It explains the working of various phases of SCM. The need for maintaining a SCM and why it is necessary to maintain a proper SCM in order to track and control each activity in the phases of SCM are also discussed in this chapter. Section 1.5 contains the features of SCM along with the advantages of SCM listed in Sect. 2.

Section 3 lists and explains various applications where SCM is maintained for better experience. Section 4 introduces to various Digital Supply Chain tools which are useful in different stages of SCM along with their functionalities. Section 5 contains the features of digital SCM as well as explains the application of artificial intelligence (AI) and blockchain in digital SCM. The last section contains a Case

Study on "Blockchain-based Drug Supply Chain Management System" where the implementation of Blockchain in SCM is discussed.

1.2 Motivation

The existing Supply Chain Management System faces many challenges such as increase in consumer demand, assuring good quality of products, in-time delivery, etc. Therefore, to overcome these challenges, evolution of SCM to Digital Supply Chain Management is needed. This can be achieved with the help of various technologies like Blockchain, Artificial Intelligence, Cloud Computing, Machine Learning, Internet of Things (IOT), etc.

The Digital SCM would completely eliminate paper-based record storage method and eliminate the manual data entries as everything will be updated automatically at each phase. Also, retrieving the data would be easy. With the help of sensors and tracking facilities it is possible to track the shipment of all goods during manufacturing, transportation and logistics processes at real-time. Quality and Transaction information can be recorded with RFID technology [1]. Various tools can be used to manage the processes with less human interaction. Resource planning, Inventory management, Process planning, etc. can be efficiently managed with digital supply Chain.

1.3 How Does SCM Work?

Supply Chain Management (SCM) is a key process for many of the companies so that their supply chain will work efficiently and effectively [2]. A company makes use of five steps to transform the unprocessed materials into completed products. Figure 1 shows the working of Supply Chain Management in a cyclic format [2]. It involves five essential modules of supply chain management. Let's see each of them in detail.

- **Planning**
  The first and the main stage of the supply chain process is the planning stage in which a plan is to be developed to verify that that the products and services which we are providing to the customer will satisfy their requirements. During this stage, the focus should be given on maximizing the profit margin.
- **Development**
  The second step consists of developing and sourcing. During this stage, the main focus is given on building a strong bond with the suppliers of the raw materials which are required for production. This stage focuses mainly on identifying dependable suppliers who will provide appropriate material for the production

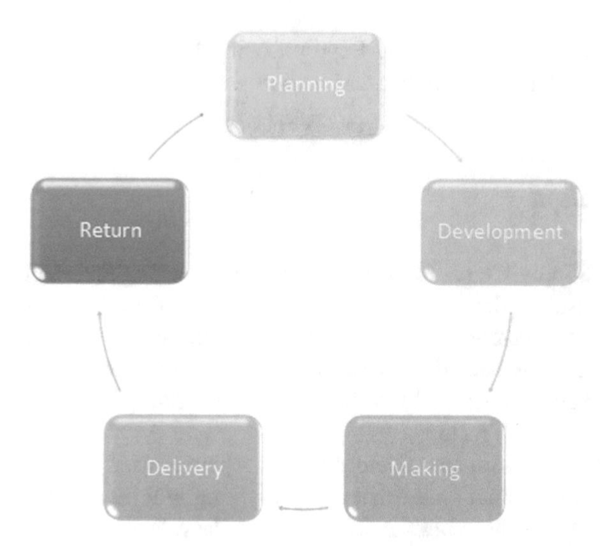

Fig. 1 Steps in SCM

purposes and as well as for determining some different planning methods for shipping, delivery and payment of the respective product.

- **Making**
  The third step consists of manufacturing and making of the products as per the demand of the customers. During this stage, the products are designed, manufactured, checked, wrapped, and synchronised for delivery.
- **Delivery**
  The fourth stage is known as the delivery stage. In this stage, the products are delivered to the customer at the given location by proper supervision. This stage is the logistics phase, where customer orders are accepted, and the delivery of the goods is planned.
- **Return**
  The final stage of supply chain management is the return stage. In this stage, the customer returns the damaged and defective goods. In this stage, the companies need to deal with queries which have been raised by the customers and as well as they have to give answer to their complaints. This stage is many times called the most troublesome stage of the supply chain for many of the companies. The planners of the supply chain have to look out for better options and solutions to handle these defective and damaged products which are received by the customers in return.

1.4 Why Is It Necessary?

There are various reasons that SCM is necessary for retailers and many similar kinds of businesses. SCM helps in boosting customer service, reducing operating charges, and for improving a company's financial position in the market [3].

- **Client Satisfaction**:
  The most important thing in any business is to know whether their clients are satisfied with them. The correct product and quantity must be delivered to the customer on time and efficiently to ensure that they are satisfied. SCM can guarantee that the customers are happy at all times, which can eventually improve the business of that respective company.
- **Reduction in Operating Charges**:
  Supply chain management decreases the gross supply chain charges as most of the manufacturers depend heavily on supply chain managers for creating networks such that customer service goals are met at the most affordable rate. Retailers also require supply chains to ensure the delivery of expensive products on time to limit inventory.
- **Improves Financial Position in the Market**:
  SCM increases profit margin since supply chain managers help to maintain and reduce the cost of the supply chain, which increases the company's profit drastically. Supply chain managers also decrease the use of plants, warehouses, and vehicles within the chain so that most of the assets will remain safe with the company. Finally, SCM can increase your business's cash flow, given that customers can receive their products faster thanks to supply chain managers.

1.5 Features of SCM

- **Connected**
  For the supply chain to work efficiently, the mutual connection between all components is essential. We can use technologies like cloud, Internet of Things (IoT), RFID, to establish this connection. Real-time data of processes carried out should be visible in a well-connected supply chain. Transparency in the supply chain provides a bigger picture which allows the managers to deploy solutions to problems without any delay.
- **Automation**
  Routine tasks should be automated so that employees can spend a significant amount of time on more critical functions which in turn will generate more revenue [4]. Organisations can involve AI and ML to automate other tedious tasks as well.
- **Secure**

SCM must be protected from cyber-attacks, viruses, malware. Blockchain is one groundbreaking technology which provides security which cannot be compromised. Antivirus software is also used to protect the system from any possible threats or data loss.

- **Scalable**
  SCM software should grow or shrink in real-time. As a business grows, its customers increase, transactions increase; this should not affect the system's performance. Allocation of hardware resources should be based on the increase or decrease in the volume of workload.
- **Analytics**
  SCM system should provide analytics tools like Demand forecast, inventory analysis, cost analysis, order processing analysis. These analysis tools help the organization in making decisions and plan further steps.
- **Fault-Tolerant**
  Backup storages should recover any loss of data. Failure in any one component of SCM should not result in crashing the whole system. Other parts should work efficiently till the problem is resolved.

2 Advantages of SCM

- **Consumer/Client Satisfaction**
  As the essential task of this management is to make sure that goods reach the client or consumer at the exact time which eventually results in consumer satisfaction, as these days people give more preference towards the quality provided by the company and supply chain management does precisely that, resulting in the client of the company getting satisfied [5]. We all know this thing that a happy customer will not only bring his business but also he will be acting as a spokesperson for our company which will eventually help in the positive growth of the business.
- **Better Collaboration**
  When two or more companies unite with each other, they share valuable information which eventually helps both the companies to grow seamlessly and because of improved access to data supply chain leaders get the information they needed in context to make more informed decisions [5].
- **Improved Quality Control**
  When a company has greater control over not only their suppliers but also with their supplier's supplier, then it is guaranteed that the company will have improved quality control over their products [5].
- **Shipping Optimization**
  The key role of any reputed company is to identify the best shipping method for handling small parcels as well as orders in bulk, for other scenarios, it helps the companies to get orders to their customers by saving time as well as money [5].

3 Applications and Use-Cases

- **Automated Shipping and Tracking**
  Logistics is a field which includes warehousing, shipping, cargo, courier services, road or rail transportation, air freight. SCM can be applied here. If a business is manufacturing and shipping their products without the help of any other firm, then this transport management system is included within the existing SCM system.
- **Manufacturing Industry**
  Any large-scale manufacturing organization has the use of SCM. All activities right from procurement of raw material to marketing or delivery of the product. SCM fosters a quality culture and enhances competitiveness in the organization. Figure 2 shows some of the components often found in manufacturing SCM [6]. Better productivity is achieved in manufacturing processes.
- **Blockchain supply chain management**
  Supply chains consist of complex networks connecting suppliers, manufacturers, transporters and consumers [7]. Shared infrastructure of blockchain proves to be a boon in supply chain management which streamlines workflows and no matter how big the size of the network becomes, blockchain can handle it.

  The third-party attacker cannot change data in the system. It improves the consumer experience by providing transparency and traceability. Supply of counterfeit goods in the market can be minimized with the help of blockchain. E.g. using blockchain, the supply of counterfeit drugs in the market can be stopped.

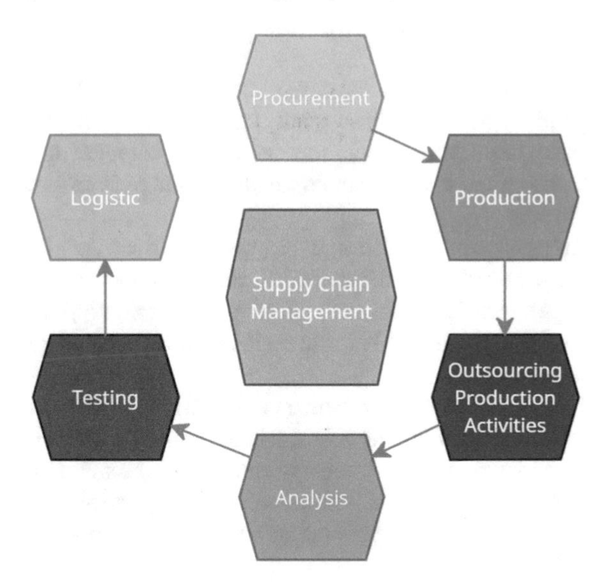

Fig. 2 Components of SCM in manufacturing

4 Types of SCM Tools

Maintaining a supply chain is a task of responsibility, and a single error may lead to a considerable loss of the company. So, if a good supply chain is to be maintained, they should use proper tools. Specialized tools make it possible to manage a supply chain effectively and reduce errors which will ultimately benefit the company [8]. Here are few SCM tools which will help companies in efficiently managing their supply chain.

- **Real-time Shipping Status Updates**
  Many companies supply their products to customers around the country and even around the world. To know the timely updates about the shipment, they require a tool to track the shipment and provide updates. A real-time alert system provides timely updates about the shipment activities [8]. It provides the updates to all stakeholders so that necessary actions can be taken in time if there's an issue. We can get these real-time alerts on our mobile devices to inform about the status of our supply chain.
- **Order Processing**
  It is an essential phase in supply chain management. It involves various tasks which can be managed effectively with the help of these tools. Such tools support all functions across order processing like order processing, order management, order fulfilment and billing [8]. These tools automate most of the activities that are involved in order processing to capture order data directly. It reduces time and errors as every task is automated.
- **Warehouse management**
  These tools will help to manage the day-to-day functions in the warehouse. It has various capabilities for warehouse management as per the company requires. Some advanced tools have capabilities to handle complex logistics [8]. The company can also use it to bundle multiple products kept at different warehouses.
- **Freight Handling**:
  Along with shipping functionalities, tools provide various freight handling functionalities as per industry requirements [8]. Some products need to be kept at a particular temperature right from the beginning till its delivery; these tools provide such functionalities too. It also helps to maintain good product quality.
- **Supplier Management**
  Maintaining good relations with suppliers is essential. These tools help in understanding the relations with suppliers. But, it is not limited to only this, by analyzing the supplier's performance, the company can figure out how much the supplier has contributed to their business. Also, based on the performance, the company can take decisions regarding supplier relationship management.
- **Analytics and reports**
  Along with analyzing supplier's performance and consumer demand, these tools provide an analytics report on the complete supply chain [8]. It gives the performance of the company as a whole or as an individual sector. Also, it uncovers the sources of delays and issues in order processing. Analysis of transportation

and logistics processes can also be done with these tools. Decision-makers can take decisions to manage various methods that are having issues by reviewing the analysis report.

- **Security features**
  The primary concern that comes into the picture when it comes to data is security. Data theft may cause the company to lose its position in the market as well as it may affect the relations with suppliers [8]. In order to prevent such security failures, the company should implement a secured network which can be achieved by ensuring that only approved personnel have access to specific company data.
- **Transportation and Logistics**
  Transportation and logistics tools assist in managing the movement of materials from one location to another [8]. It helps in planning and tracking of the shipment and if there is an issue, takes necessary actions accordingly.

5 SCM via Digitalization and Artificial Intelligence

As discussed earlier, the supply chain involves various tasks like tracking shipment, order processing, warehouse management, analysis, security and many more, managing all these tasks is quite difficult. Therefore, there is a need to take control of these tasks to the next level with the help of some technology that can simplify the workload. One solution to this is to automate these processes by using Artificial Intelligence. Another is to maintain a secured supply chain by using blockchain technology. Also, Computer Vision can be used in quality inspection of various food products [9]. RFID technology can be used in identifying counterfeit products that enter into market [10, 11]. These technologies will help in many aspects like quality management, security, analysis, decision-making process, etc. It has been discussed in brief in the following sections.

5.1 Features of Digital Supply Chain

- We can easily connect and relate data from various sources.
- The generation of data-driven plans via data visualization is also one of the advantages of digital supply chain management.
- One of the most significant advantages is that it provides automation, which eventually increases the efficiency of the system or the platform by eliminating the manual checking for errors and increases the accuracy.
- One more important characteristic of digital supply chain management is a collaboration via which we can connect multiple internal and external systems and people.
- Another advantage of digital supply chain management is data analytics; when such a tremendous amount of data is converted to digital data, one can

quickly identify bottlenecks, point out savings, various patterns and opportunities regarding cost reduction.

5.2 Artificial Intelligence in SCM

5.2.1 Definition of Artificial Intelligence (AI)

Artificial Intelligence can be defined as the technology where machines are programmed and trained to think like humans and perform actions accordingly.

- Some examples of AI in our day-to-day life are as follows:
- Google Maps
- Voice Assistants like Alexa
- Social Media
- Face Detection
- Spell Check or Autocorrect
- Chatbots
- E-Payment
- Digital Assistants.

5.2.2 Digitalization of AI in SCM

Artificial Intelligence (AI) and Machine Learning (ML) have changed the world completely. It has brought everything on our fingertips, and with their help, each and everything has become very convenient. AI in supply chains is helping a lot to deliver the powerful optimization capabilities required for more accurate planning, high quality, better customer satisfaction, lower costs and greater output [12]. AI methodologies can be used in Supply Chain Risk Management (SCRM) [13].

5.3 Applications of AI in SCM Activities

5.3.1 Chatbots for Operational Procurement

A chatbot is a program which simulates or translates human conversation through voice commands or text messages [12]. This requires access to robust and intelligent data sets and so that AI plays a significant role over here. For daily tasks, Chatbots can be utilised for:

- Speaking with suppliers during trivial conversations
- For setting as well as sending actions to the suppliers regarding governance and compliance materials

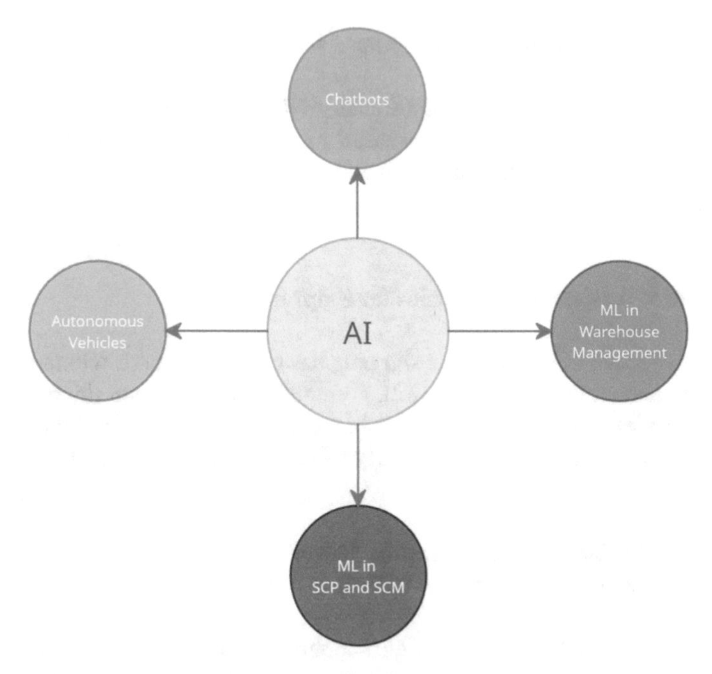

Fig. 3 Applications of artificial intelligence in SCM

- For the placing purchasing requests
- Receiving or filling of invoices and payments or order requests.

Figure 3 shows the different areas of the supply chain, where Artificial Intelligence can be used to simplify the tasks.

5.3.2 For Enhancing the Experience of Customer

By the use of AI and ML, the relationship between the buyers and the suppliers has got much better [12]. The best example of it is Amazon Alexa which helps the customers to track their respective order.

5.3.3 Machine Learning for Supply Chain Planning

The supply chain is an important activity within Supply Chain Management strategy. In today's business world, there is a need for smart and intelligent tools. When it is applied with Supply Chain Planning, it could help in forecasting within the inventory, demand and supply. If ML is used correctly through SCM work tools, then it could revolutionize the agility and optimization of supply chain decision making.

5.3.4 Machine Learning for Warehouse Management

After having a closer look towards the domain of SCP, its success is heavily dependant on proper warehouse and inventory-based management [12]. But if these things are ignored, then it can prove a massive disaster for the company because proper warehouse management is essential.

5.3.5 Use of Autonomous Vehicles for Shipping

Use of Intelligence in logistics and shipping has gained the focus within the supply chain management in recent years [12]. Faster and more accurate shipping reduces transportation expenses, reduction in labor cost, it also adds more environment-friendly operations, and more importantly widens the gap between the competitors.

5.3.6 Machine Learning for Supplier Selection and Supplier Relationship Management (SRM)

Supplier Selection and selecting the right supplier for sourcing is crucial for a sustainable supply chain [12]. If at all anything goes wrong while choosing the supplier, then the entire business can suffer. Machine Learning and Intelligent Algorithms can help in predicting the right supplier selection and risk management, during every single supplier interaction, by analyzing data sets generated from SRM actions. With the help of this technique, we can use this data for the betterment of the business.

6 Supply Chain Management Using Blockchain

Blockchain has emerged as a secured network in financial services and is gaining considerable attention for addressing supply chain management issues. Due to the entry of counterfeit products into the market, the importance of the quality of products and management of the supply chain has increased. Blockchain has characteristics such as decentralization, security, immutability and smart contracts that can bring significant benefits if used properly [14]. Blockchain can also be used to manage the transport and logistics processes [15].

6.1 Decentralization

It is different from the traditional transactions that need to be approved by central authorities [14]. Decentralization eliminates the central powers and allows direct transaction between users. All details can be recorded about the product, suppliers, and customers along the whole supply chain. Multiple stakeholders have a copy of records which can be retrieved anytime. Producers can also make sure that the raw material quality meets requirements.

6.2 Security

Due to the decentralization feature, failure of a single node will not lead to the failure of the whole network, which can reduce the chance of hacking [14]. When applied to the supply chain, blockchain can keep data safe, reducing the risk of hacking and data stealing. If any user wants to add or update a block, it is first verified and approved by other members, and then the transaction is performed.

6.3 Immutability

Along with data security, it also ensures the record's originality and authenticity [14]. This means historical data cannot be changed without informing other members. It allows companies to trace back along the supply chain to identify if there are any issues and take actions accordingly.

6.4 Smart Contract

It is another essential feature of blockchain, which is a digitalized contract and operates once certain agreements are met [14]. It can speed up the transactions and enhance trust. For example; automatic payments can be made once the materials are delivered successfully. It can save paperwork and also minimize errors as compared to the traditional supply chain.

By implementing all these features in the supply chain, we can achieve maximum digitalization. It improves efficiency, saves processing time and provides security. Figure 4 shows the interconnected network of supply chain services using blockchain technology [16].

7 Benefits of Using Blockchain in SCM

Benefits of using blockchain technology in SCM are:

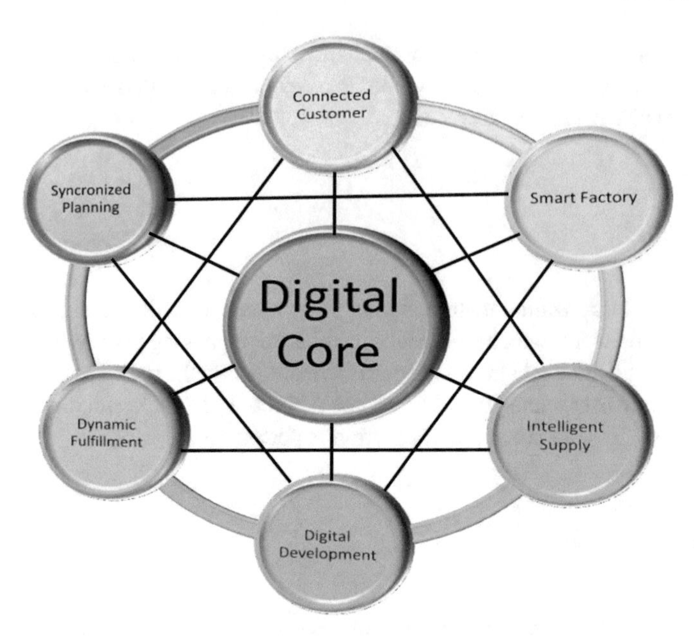

Fig. 4 A supply chain integrated using blockchain [16]

7.1 Traceability

It can be defined as the ability to trace back along the supply chain of a product when required [17]. It helps in checking the safety and quality of the product at any time. If there are any issues in the supply chain, it can be traced easily by referring to the information stored during each phase.

7.2 Digital Identity

Blockchain networks can store various types of information and provide a digital identity. It will provide unique identity and security to products. Hence, it will work as a primary construct for stakeholders in the supply chain [17].

7.3 Anti-counterfeiting

The supply of counterfeit products into the market is increasing, which affects both customers as well as the company. Hence, we must use a proper mechanism to identify the original products. By scanning the barcode, all information about the product can be viewed. With this, the supply of counterfeit products can be reduced.

8 Case Study: Drug Supply Chain Management Using Blockchain

The emergence of Counterfeit drugs in the market cause tremendous loss to the pharmaceutical industry and the government [18]. An enormous amount of money is spent by the government to put a check on these counterfeit drugs. These fake drugs do not heal the patient but may have other dangerous side effects. It is difficult to identify and separate fake medicines as they cannot be segregated through their appearance. Sophisticated labs are required to test the authenticity of the drug. So, it isn't easy to separate fake medicines once they are in the market with authentic other medicines. The only way to make sure that counterfeit medicines do not enter the market is to track the medicines from the beginning of their supply chain, and only those medicines should be sold which are from a well-known licensed company. One of the possible solutions that can be proposed to this problem is to track the delivery of drugs in each phase right from raw material procurement from supplier, manufacturing stage, distribution stage, pharmacies, and clinics and to consumers finally [19].

8.1 Proposed Solution

Blockchain-based Drug Supply Chain Management system can be maintained using the smart contracts to track the products in each phase.

8.2 Workflow of the System

The diagram that is shown in Fig. 5 depicts that the users (supplier, manufacturer, distributor, doctors and pharmacies) can manage the whole system. Each user can perform transactions on the Blockchain network through a web-based application provided to them, which acts as an interface between the users and the network. All members of the network can view and update the status of the system and the data stored in it. Whenever an action takes place, like an authorized supplier supplied raw materials to manufacturer; a transaction in blockchain is issued by the supplier that he has supplied the raw materials. On receiving this raw material, Manufacturer will confirm that the consignment is received. This same process will take place between manufacturer-distributer and distributer-pharmacies/hospitals. In this way drugs from non-authorized suppliers can be prohibited. Patients, on the other hand, can stay relaxed as they can track the source of the medicine, its price, expiry date, manufactured by, etc. they are consuming. The medical staff may not use visualization of data in the system, but academicians might use it for their projects, and research.

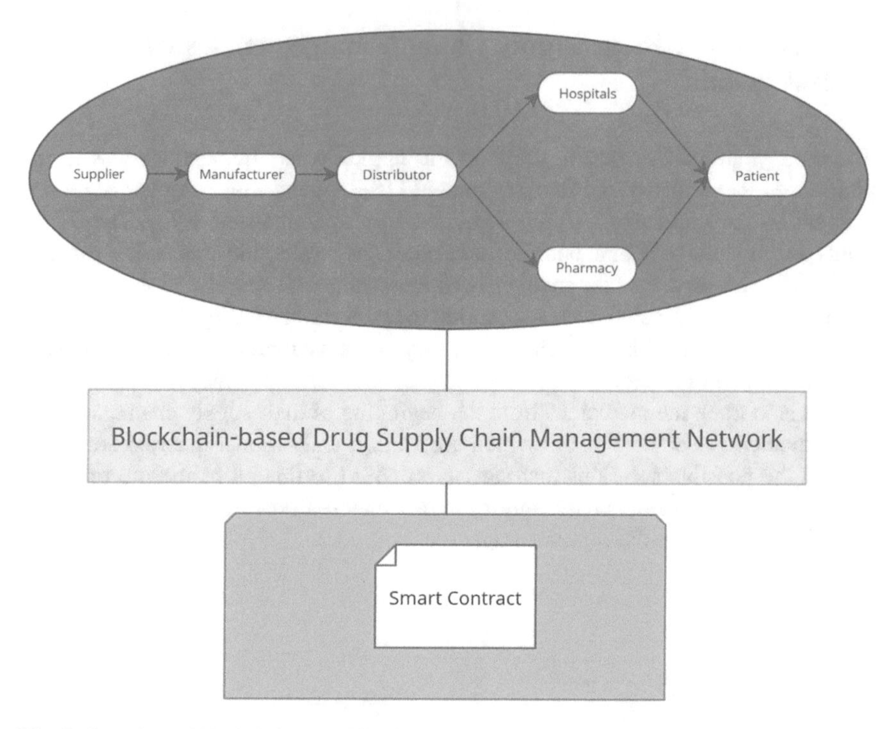

Fig. 5 Overview of blockchain-based DSCM system

Let's consider a scenario: A manufacturer needs some raw material to manufacture medicines. He then places an order for raw-material to the supplier. The supplier validates his transaction, and it is added in the blockchain. Similarly, when the supplier supplies the raw materials to manufacturer he adds the details like material name, quantity, cost, etc. into the system. Similar kind of procedure takes place when manufacturer supplies the drugs to distributor and adds the manufacturing details like manufactured by, date of manufacturing, date of expiry, etc. as shown in Fig. 6. Same procedure is followed by the distributors, hospitals and pharmacies. As all transactions are stored in the Blockchain, they can be tracked during the process and after the process as well.

Let us consider another scenario as shown in Fig. 7 where manufacturer requests raw materials to the supplier by adding the details into the system. Similarly, when supplier supplies the requested raw materials to manufacturer respective block will be added into the system.

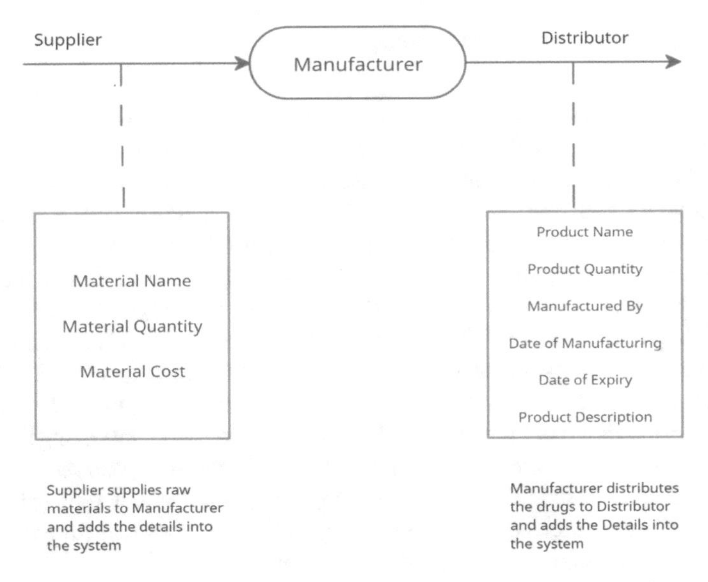

Fig. 6 A scenario of manufacturer receiving raw materials and adding manufacturing details after distribution

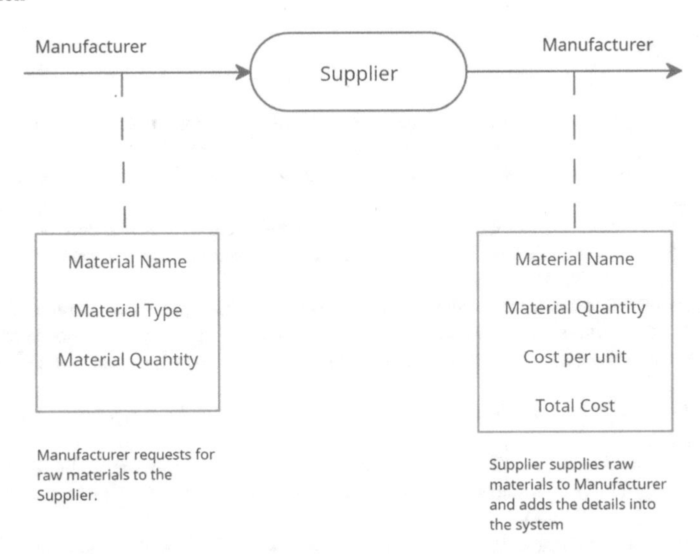

Fig. 7 A scenario of manufacturer requesting raw materials to Supplier and receiving it

9 Conclusion and Future Scope

Various technologies can be used for tracking and analysis purpose such as Machine Learning, Natural Language Processing, Blockchain, IoT and so on. With the help of these technologies a secure and efficient SCM can be maintained. Blockchain-based Drugs Supply Chain Management can be used to track and maintain each and every phase of a SCM efficiently. It will also help in identifying the fake drugs that enter into market. Therefore, the emergence of fake drugs into market can be reduced to a greater extent. This system is not only limited to only Drugs SCM but, can also be used in various types of SCM systems such as Food Supply Chain Management where the product's quality is main objective and hence a proper SCM should be maintained. With the ability of Traceability organization can trace back through each phase and figure out the problem in case of any issue occurs during the further processes. Hence, to overcome the challenges in existing SCM, one solution can be Digitalized or AI-based SCM. In future, along with these technologies IoT can also be used for better efficiency where RFID tags can be used for real-time tracking of shipments and sensors can be used of quality management.

References

1. Chen, S., Shi, R., Ren, Z., Yan, J., Shi, A., Zhang, J.: A blockchain-based supply chain quality management framework (2017). https://doi.org/10.1109/ICEBE.2017.34.
2. https://www.tutorialspoint.com/supply_chain_management/supply_chain_management_process.html
3. https://www.yourarticlelibrary.com/retailing/why-scm-is-required-5-needs/48284
4. https://www.zoho.com/blog/creator/8-essential-features-of-an-effective-supply-chain-management-solution-by-david-rogers.html
5. https://www.letslearnfinance.com/advantages-disadvantages-supply-chain-management.html
6. https://www.business-software.com/article/key-features-of-supply-chain-management/
7. https://www.supplychaindigital.com/supply-chain/how-ensure-truly-connected-end-end-supply-chain
8. https://www.selecthub.com/supply-chain-management/13-essential-supply-chain-management-tools/
9. Patel, K., Kar, A., Jha, S., Khan, M.: Machine vision system: A tool for quality inspection of food and agricultural products. J. Food Sci. Technol. **49**, 123–141 (2012). https://doi.org/10.1007/s13197-011-0321-4
10. Chowdhary, C.L.: Growth of financial transaction toward bitcoin and blockchain technology. In: Bitcoin and blockchain, pp. 79–97. CRC Press (2020)
11. Coustasse, A., Arvidson, C., Rutsohn, P.: Pharmaceutical counterfeiting and the RFID technology intervention. J. Hosp. Mark. Public Relat. **20**, 100–115 (2010). https://doi.org/10.1080/15390942.2010.493369
12. https://www.manipalprolearn.com/blog/6-ways-ai-making-supply-chain-more-seamless-supply-chain-aka-logistic-industry
13. Baryannis, G., Validi, S., Dani, S., Antoniou, G.: Supply chain risk management and artificial intelligence: state of the art and future research directions. Int. J. Prod. Res. (2018). https://doi.org/10.1080/00207543.2018.1530476

14. Duan, J., Zhang, C., Gong, Y., Brown, S., Li, Z.: A content-analysis based literature review in blockchain adoption within food supply chain. Int. J. Environ. Res. Public Health **17**(5), 1784 (2020). https://doi.org/10.3390/ijerph17051784
15. Pournader, M.: Blockchain applications in supply chains, transport and logistics: a systematic review of the literature. Int. J. Prod. Res. **58** (2019). https://doi.org/10.1080/00207543.2019.1650976
16. Mukri, B.: Blockchain technology in supply chain management: a review (2018)
17. Paliwal, V., Chandra, S., Sharma, S.: Blockchain technology for sustainable supply chain management: a systematic literature review and a classification framework. Sustainability **12**(18), 7638 (2020). https://doi.org/10.3390/su12187638
18. Jamil, F., Hang, L., Kim, K., Kim, D.: A novel medical blockchain model for drug supply chain integrity management in a smart hospital. Electronics **8**(5), 505 (2019). https://doi.org/10.3390/electronics8050505
19. Abbas, K., Afaq, M., Ahmed Khan, T., Song, W.-C.: A blockchain and machine learning-based drug supply chain management and recommendation system for smart pharmaceutical industry. Electronics **9**(5), 852 (2020). https://doi.org/10.3390/electronics9050852

The Role of Machine Learning Techniques in SCM—An Analysis

B. K. Tripathy, S. Parikh, R. Jhanwar, and P. Ajay

Abstract Supply chain is a network of organizations that includes retailers, wholesalers, carriers, producers, and clients. It should work firmly to increase the customer value. It necessitates that the entities should coordinate to share data and integrate them. However, the real universe and the ideal one for supply chain networks differ from each other. This is a result of the existence of known and unknown elements, ideas inherent in supply chain. Utilization of merchandise, dating back to the 2000s and now are two different aspects as the demand of customers has increased exponentially. For the market to satisfy the need, we require an effective strategy for the supply chain. Subsequently, Machine Learning became an integral factor in dissecting and providing ideal answers to overcome this supply and demand issue. In this chapter, we shall provide an outline of machine learning along with interpretation of some models under it, highlight their utility in real life and focus on how ML is useful in Supply Chain Management. We shall analyze its different aspects and parameters, with the help of a case study and its thorough analysis. This will provide a detailed insight into all the advantages and disadvantages of using ML in SCM, and its prospects in the future.

Keywords Supply chain management · Machine learning · Forecasting · Predictive analytic

B. K. Tripathy (✉) · S. Parikh · R. Jhanwar
School of Information Technology and Engineering, VIT, Vellore, India
e-mail: tripathybk@vit.ac.in

P. Ajay
School of Electronic Science and Engineering, VIT, Vellore, India

K. Perumal et al. (eds.), *Innovative Supply Chain Management via Digitalization and Artificial Intelligence*, Studies in Systems, Decision and Control 424,
https://doi.org/10.1007/978-981-19-0240-6_2

1 Introduction

In today's world as we thrive to achieve automation over increasingly complex industrial processes, perceptive and efficient logistical processes play a major role. The foundation of a successful supply chain is the combination of High logistic requirements such as reliability, transparency and Flexibility and optimal economic conditions [43]. As predicted in [31], the size of the market is likely to go up to USD 8.8 billion by 2025, which stands at USD 3.5 billion now. During this period the forecast comes out to a Compound Annual Growth Rate (CAGR) of 19.8%. Such a huge market deals with large volumes of data and dynamically changing operations which makes the process of decision making robust. Hence a technology that is able to deal with the increasing complexity of supply chains.

In this intrinsic and volatile network of supply chain, there evolves a need for meeting the new demands within maintaining a smooth sailing flow. Supply Chain Management (SCM) hence falls as a prey to challenges given below due to its operational inefficiencies:

- Demand Variations
- Supply Shortages
- Poor communication within the SCM
- Backlogs of orders
- Uncertainties in logistics
- Inventory mismanagement
- Inconsistency quality of the inventory
- Variations in the performance indicators used by the customer.

Reasons for these gaps [17] could be "alignment of business interests, long term relationship management, reluctance to share information, complexity of large scale supply chain management, competence of personnel supporting supply chain management, performance management and incentive systems to support supply chain management". The prime contributors to these challenges differ on a daily basis and it is difficult to forecast using known techniques. Techniques of Machine Learning (ML) provide a more efficient and realistic forecast than its traditional counter parts [10].

Machine learning has revolutionized industries to great extents by optimizing their routine procedures. Pattern identification in the complex supply chains has the perspective of revolutionizing businesses. Machine learning algorithms are sustained enough to perform this job daily on huge datasets evolving from supply chains, with no support from human beings and the analysis does not require any taxonomy being explained. Identification of primary set of parameters impacting the process is carried out with great accuracy, using modeling with constraints through iterative querying of data by the algorithms. Based on which the key elements influencing demand forecasting, order-to-cash, procure-to-pay, delivery management, inventory volumes, supplier quality, planning production [13] and more are encountered.

2 Machine Learning

ML deals with computer algorithms which follow the continuous learning process to improve themselves. Coming under AI, it takes care of applications which process without being coded explicitly for the purpose to predict outcomes with good accuracy. Outcomes are predicted using existing data by ML algorithms showcases the percentage of particular tasks in ML algorithms (Fig. 1).

2.1 ML Types and Their Interpretation

Basing upon the processes used by the algorithms to learn and predict, ML is categorized into three broad categories (Fig. 2). These are supervised learning, unsupervised learning and reinforcement learning. Supervised learning covers two types of algorithms; classification and regression, whereas unsupervised learning covers two more

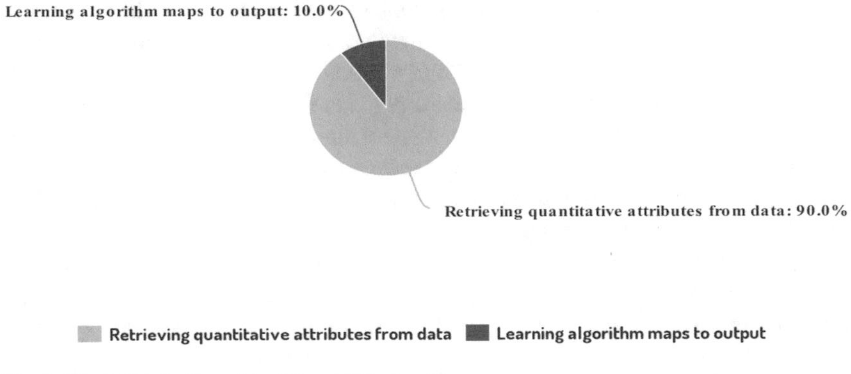

Fig. 1 Work distribution of traditional ML algorithms

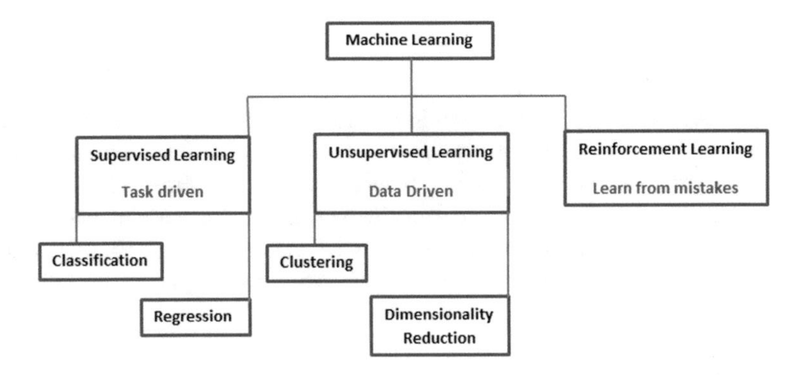

Fig. 2 Types of machine learning

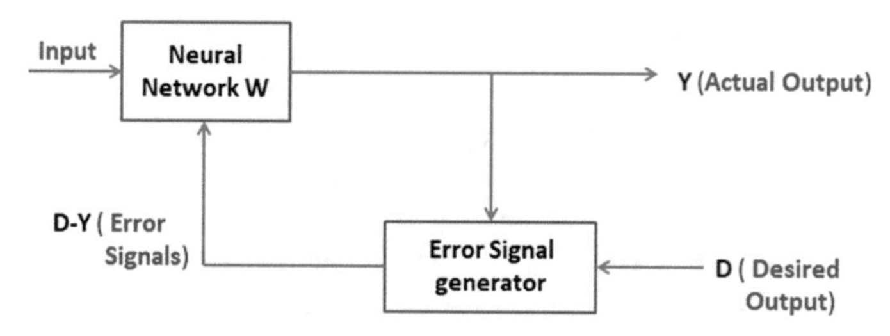

Fig. 3 Supervised learning

types; clustering and dimensionality reduction. Depending upon the application and its requirement a data scientist selects an appropriate algorithm for prediction.

2.1.1 Supervised Learning

It is one of the basic and widely used kinds of ML. The algorithms under this category are trained by taking some labeled data. The data available are divided into two parts; training data and testing data. Although there are no specific guidelines regarding this partition, it is suggested that the size of training dataset must be higher than the size of the testing dataset. It is a standard practice that the ratio may be 60:40 or 70:30 or 80:20. These are powerful and efficient algorithms under conditions which are favourable. The training and testing phases are characteristically different from their roles point of view. In this type of learning basically a function is learnt which is done under the given input output pairs. The output value for a given input value is called the supervising signal. At every step of the iteration, the difference between the actual and desired output, called the error, is determined and fed back to adjust the values of the parameters involved. At the end of the training phase the system is ready and is tested for its accuracy by using the testing dataset. The number of test cases correctly identified determines its accuracy (Fig. 3).

2.1.2 Unsupervised Learning

Unlike supervised learning, in unsupervised learning the dataset is not labeled. There is no human intervention in this process, which requires larger and irregular datasets. The training process ends up by finding hidden irregular structures. This makes the unsupervised learning methods more versatile than supervised learning approaches. Similarity among the data elements is used to find their similarity to put under groups. The actual output is put into the system back for adjusting the parameters (Fig. 4).

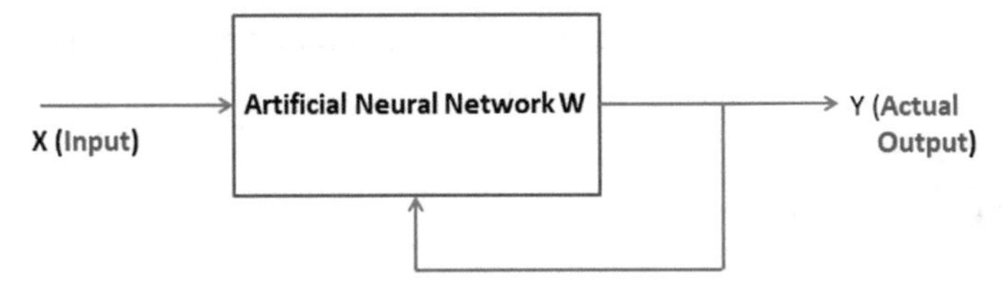

Fig. 4 Unsupervised learning

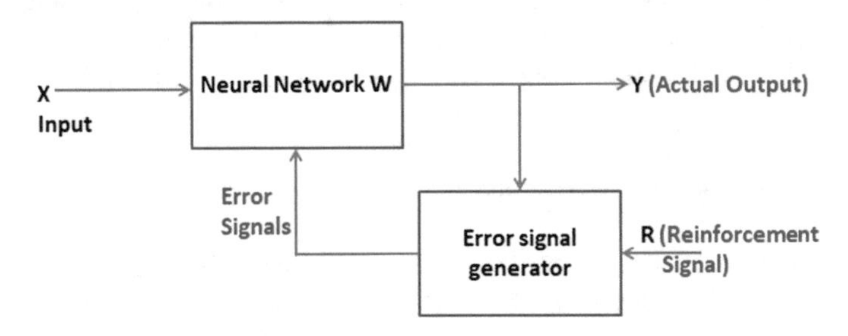

Fig. 5 Reinforcement learning

2.1.3 Reinforcement Learning

Reinforcement learning takes place basing upon critic's information. There is no labeled data either. So, the error signals are generated depending upon the critic's observation and error signals are generated (Fig. 5). These error signals are fed back to adjust the parameters. Thus the system enhances itself. Mostly trial and error methodology is adopted. The critic basically behaves as an interpreter. The interpreter decides whether the output is acceptable or not at the end of each cycle.

2.2 Machine Learning Algorithms

As mentioned above several machine learning algorithms exist in literature. In this section, we present some of these well-known approaches.

2.2.1 Linear Regression

A linear relationship between a dependent variable and one or more independent variable is called linear regression. It may be simple or multiple depending on whether the number of independent variables is one or more. The graphical representation of such a model is always a linear graph. Simple regression has the equation of the form $y = ax + b$, in the slope and intercept form of a straight line. Linear regression can be used to analyze the marketing effectiveness, promotions and pricing on sales of a particular product.

2.2.2 Decision Tree

This model uses the approach of supervised learning algorithm because of its use case in classification related problems. Here, the dataset is split into two or more equivalent sets. This is achieved using the most crucial attributes of the independent variables in order to develop a variety of groups. The decision tree is used in real life in various sectors like engineering, civil planning, law, and business.

2.2.3 SVM (Support Vector Machine)

This is another model that uses supervised learning approach to handle classification problems. Here, all the data points are plotted on a n-dimensional space (n = number of features) via assigning the value of each feature and getting the value of respective coordinate. SVM model is used for identifying and classification of genes, patients on the basis of genetics and other biological problems.

2.2.4 Naive Bayes

This is also a classification model like SVM which is built on the known Bayes' theorem with an independent assumption of predictors. This model assumes that the existence of any unique attribute in a class is completely unrelated to the presence of any other attribute. The use of this classifier lies in problems having numerous classes and text classification. Kernel density estimation is mostly used for conditional probability estimation and it is a key problem in naïve Bayes models [23].

2.2.5 Neural Networks

A neural network is a model which comprises of several neurons arranged in vertical columns called layers such that there are connections directed from each neuron to all the neurons in the next layer. The first layer is called the input layer and the last layer is called the output layer. Each of the connection is attached with a weight, which contain information about the input signals. In order to compute the output at a neuron, functions called activation functions are applied on the total input. Figure 6 provides architecture of a multilayered neural network.

There may be several intermediate layers called the hidden layers. But, increasing the number of hidden layers increases the complexity of the network. The function and structure of neural networks are modeled after the animal brain. The neural networks are capable of modeling complex functions from given datasets. We shall

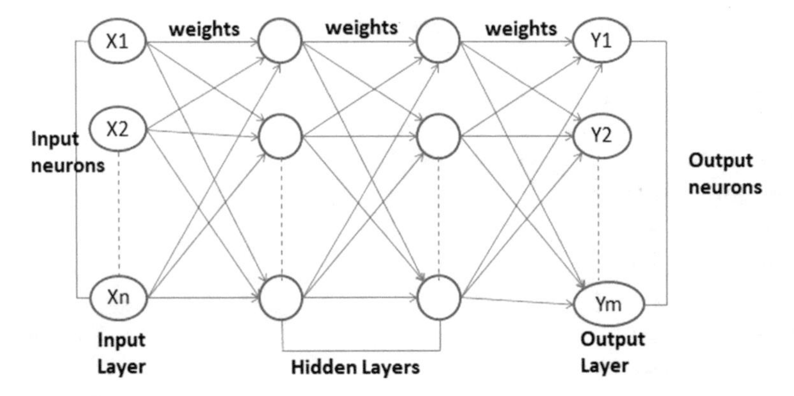

Fig. 6 Architecture of a multilayered neural network

discuss about the latest developments in neural networks under the deep neural networks in a later section.

3 SCM Now and a Decade Ago

Over the previous decade or two, the history of SCM has progressed from the initial simple but labourious processes into complex ones involving broader strategies and management of materials in a distributed environment which are extensive enough to even cover the entire world.

In this section, we provide more information about SCM and the elements that have impacted it during the previous decades.

Different meanings of SCM have been offered in the course of the last 20–30 years; however, none of these definitions is general. The absence of an all-inclusive meaning of SCM is to some extent because of the manner in which the concept of the supply chain originated [14]. The idea of the supply chain has been considered from various perspectives in various assemblages of writing. Such multidisciplinary beginnings make it hard to think of a general meaning of SCM.

According to [8], the concept of SCM is not new and is obtained as a suitable combination of purchasing and distribution functions. The evolution of SCM [8] is shown in Fig. 7. The evolution of SCM has been categorized into the following three phases.

- Activity fragmentation (1950s and 1960s)
- Activity integration (1960s to 2000)
- Supply chain management (2000 onwards).

The idea of SCM was blurred in the 5th and 6th decade of the twentieth century. There were moderate advancement in fresh products and was a part of the technology and was limited by the capacity of the firm. Stock padding was hindering tasks to

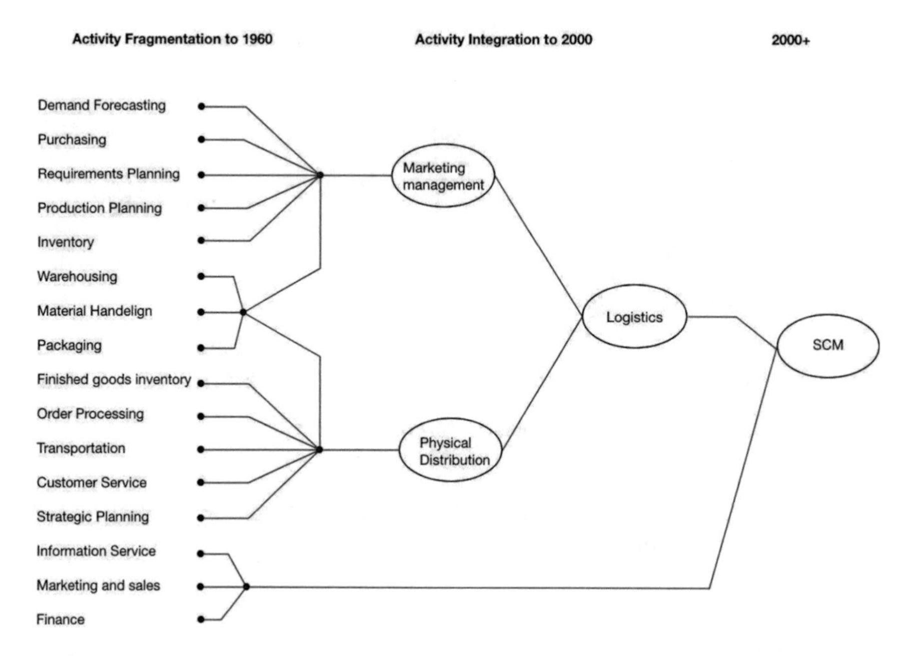

Fig. 7 Evolution of SCM

keep a reasonable base line, bringing about immense interest in work in cycle (WIP) stock as per [40]. Moreover, buying being considered as a production service was not acceptable [18]. Enhancing production was the main objective during this period. Association between a purchaser and a provider was given less importance. The information sharing in the form of technology or expertise was considered as risky and hence was unacceptable [40],

During the 1980s and mid-1990s, firms dealt with expanded requests for "better, quicker, less expensive strategic assistance". This led to focus of majority of producers being shifted towards development of core competencies and logistics activities being outsourced [15]. The external expert introduced monetarily feasible methods for accomplishing profitability and proficiency [15]. Accordingly, most of the manufacturers preferred a methodology of establishing relationship with their provider and client. This is because of their comprehension of the advantages of a helpful relationship with different firms in diverse chain levels [41].

There was a boost to the advancement of the SCM and the relationship between a purchaser and a provider in the 1990s due to the introduction of Enterprise Resource Planning (ERP). While Electronic Data Interchange (EDI) frameworks dealt with inter–organizational integration, ERP frameworks were mainly concerned with intra-hierarchical combinations [35]. SCM continues to advance in the twenty-first century with the improvement of more modern IT frameworks (web-based arrangement frameworks), which are concerned with authoritative reconciliation as well as intra-hierarchical coordination.

The most recent pattern of advancement in SCM is the development of supplier relations across nations, and all around the world [36]. Green Supply Chain Management (GSCM) is the most recent idea acquainted with the writing of SCM. Firms have grown exponentially from their initial state, and have accomplished economies of scale. With the foundation of exchange progression strategies, they are internationalizing their organizations to locate the least sources of capital and growing markets to sell their items. The traditional idea of SCM is not sufficient for being proficient in the new, hence new ideas and executives procedures (such as GSCM) are arising. An incorporated supply chain gives a significant upper hand to the individual entertainers taking an interest in the chain. Firm–firm rivalries have become chain–chain rivalries in the newly created economies [6, 25, 28, 34].

3.1 Factors Affecting the Evolution of SCM

SCM is filled with turbulence in its due course due to various factors being involved like Major Mergers and Acquisitions, Booming e-Commerce, Return to Regional Demand, Software Shift, etc. More insights on these factors are given below.

3.1.1 Major Mergers and Acquisitions

In 2000, 20% of the business was controlled by the main 50 3PLs. The number got reduced to half by 2012. Today, in practice 70% of the market is maintained by the best 50 3PLs. Exhibiting the business keeps on combining at an always expanding rate. Due to more mergers, customers could be potentially provided with higher and variety of services by the providers. However, simply having a bigger en set-up of administrations without having capabilities set up could make a likely loss of center in individual regions.

3.1.2 Booming e-Commerce

E-Commerce has existed since the availability of the internet to the public, but the number of consumers have increased dramatically over the last five years, more and more e-Commerce are coming up by the day, a research study has shown that 80% of Americans use e-Commerce to make their purchases compared to that of 20% back in 2000.

3.1.3 Return to Regional Demand

Near-shoring and on-shoring (outsourcing business processes to a nearby country or doing them locally) are practices that have gained popularity in the recent past.

Promotions (such as Make In India) that focus more on self-consumption could be the reason for this. Due to the Make In India scheme, a larger fraction of India's production is going towards Indian demand, and less towards export.

3.1.4 Software Shift

SCM, in modern times, must concurrently accommodate increasing customer demands as well as global competition. Value systems have evolved through integration of supply chains due to advancements in Information Technology (IT). Businesses have integrated IT applications in their day to day routines such as using cloud for better inventory management, tracking of order and delivery, collaboration with partners, and to save companies essential documents, among others (Ref. [30]). A sustainable competitive advantage could be established due to the crucial role played by IT practices and technique is now a confirmed fact. Therefore, escalating IT investment has potential to increase efficiency, efficacy, and profitability.

3.2 Role of Machine Learning Techniques

Enormous amount of data are generated by the supply chains which are complex. Optimal utilization and analysis of data is difficult when data is worked on manually. Making appropriate deductions and finding the right inferences is also complicated as a result of this. ML techniques solve this problem so that companies can enhance their SCM by putting their entire data into them and so can increase optimality.

4 Machine Learning in Supply Chain Management

The variations in supply chain technology is at an exceptional rate and the chances and scope which were unimaginable a few years back have now become distinct possibilities. The primary goal of SCM is to achieve Cost efficient fulfillment for customer demands and inexpensive products. The SCM network faces many hurdles among which supply risk and demand uncertainties are driving forces to major enigmas.

The five core SCM functionalities (Fig. 8) are; Demand and Supply planning, Procurement, Manufacturing, Inventory and Storage, Transportation and Distribution. Below, we discuss each one of these in detail.

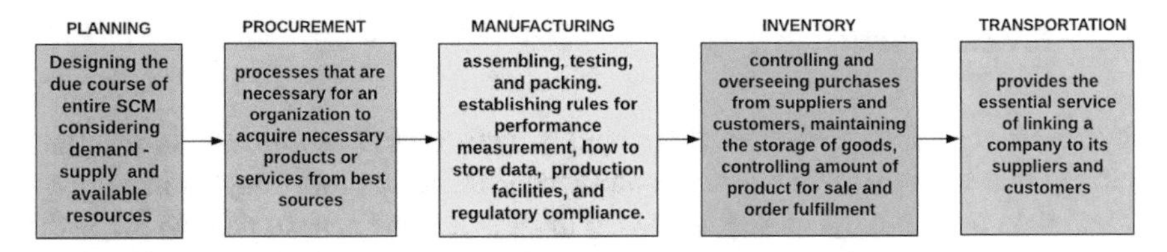

Fig. 8 Components of SCM

4.1 Demand and Supply Planning

The initial most but highly valued step in the entire SCM network is planning. Any businesses with smart planning strategies tend to have impressive turnovers at the end of the tenure. Good planning comes with experience and cleaver decision making. Evaluating from past experience is a lot important for this phase. Herewith the SCM is entirely defined for the business, i.e. the evaluation of demand and preparation of satisfactory supplies. Integrating ML with such a core of SCM is a prime choice.

Evaluation of demand for the business is a huge requirement as the following departments rely on this; it is also called demand forecasting. Recognition of distinct patterns is carried out in ML through sophisticated mathematical algorithms. Also, such types of algorithms are used to find out intricate relationships and deriving demands in diverse and big datasets. Through construction of smart systems iterative training of models and making them adaptable to changing scenarios has become reality, in addition to analyzing voluminous data and information through them. These advantages of software based upon ML techniques enhance their abilities for development of better and more reliable forecasting models which are more complex than the average ones.

The quantity and quality of some seasonal products sometimes get affected by changes in weather leading to substantial changes in their demand. Algorithms based on ML techniques are helpful in determining and measuring these impacts. Through analysis of predictions, changes can be brought in models, which support such scenarios. Changes in weather might lead to drop in demands and both the retailers and suppliers should be prepared for such unexpected changes, which are addressed effectively and efficiently through mutual coordination by planning and using the above ML techniques. It has been observed through forecasting approaches that prediction of such unforeseen effects of weather improves between 5 and 15% for every food product and this percentage as a whole goes up to 40%.

The companies' marketing strategies are quite necessary for its future. Traditional promotional methods fail to generate noticeable demand as per [5]. ML can trigger the promotions to only specific users over various platforms using sentimental analysis in order to identify potential customers for the promotion and not waste money on rest. From [5], a dairy giant witnessed approximately 20% reduction in the promotion forecast errors along with around 30% decrease in the lost sales using this strategy.

4.2 Procurement

Procurement is one department that could benefit from advancements in ML, and namely due to the many repetitive and manual tasks that most buyers have to go through on a daily basis. Tasks in procurement generate heaps of data which is the core of ML. Using ML in procurement, this department can make smart decisions, improve operations, automate tasks etc.

The processes within purchase-to-pay consist of activities of purchasing, receiving, requesting, paying and accounting for all goods and services. Use of ML under this branch has huge potential. A smart decision making model can be built to help the employees in purchasing through a catalogue. ML shall also be used to detect fraudulent transactions or to handle any anomalies in the invoices.

Smart detection models can be used in order to choose suppliers based on the multiple parameters necessary to make the best choice. ML algorithms have the potential to handle large datasets. Such results, if performed manually, might have taken months but with this technology we can obtain it within hours. The decision for righteous suppliers shall not be taken from their present but also from the series of past events in order to ensure any abnormality and genuinity of the source. Hence, using this approach, organizations can identify suppliers that have the highest possibility to bring advantages. It is evident from [10] that fuzzy neural networks are quite appropriate for the whole supplier selection process.

Verifying and analyzing contracts is the most severe and complicated job under this department. Terms and conditions shall be evaluated carefully in order to eliminate possible risks for the buyers. ML plays an important role in performing the given task. It shall also be used in order to suggest any changes to the clauses.

4.3 Manufacturing

Modern manufacturing technology has started to integrate ML algorithms throughout the complete production process; the use of these algorithms is to plan machine maintenance adaptively rather than planning on a fixed schedule.

For companies which do not have the right to develop products by themselves the manufacturing process may be costlier and take more time. As per estimation [32] the smart manufacturing markets are likely to reach 384.8 USD billion by 2025 from 214.7 USD billion in 2020. An intelligent manufacturing strategy will bring revolutionary changes to the manufacturing sector.

Production operation and maintenance of equipment contributes to a major part of expenses of those manufacturing industries which are dependent on assets. Nearly $50 billion was unplanned and 42% of it was due to major asset failure [11]. So, predictive maintenance has become a vital solution that will help to save a lot of expenditure. Trustworthy predictions regarding the status of machinery and assets using ML and neural networks has become an important aspect. Remaining Useful

Life (RUL) of equipment has increased considerably as a result. Replacement or repair of components of an equipment requires expert knowledge of the technicians and in some cases the owners are also needed to have an idea about the methods required.

It has become increasingly difficult to maintain high standards in products and controlling their efficiency and quality due to their higher complexity and short, regular deadlines. Customers expect robust and durable products. On the other hand, defective products will be required to be replaced or returned which in turn is very much likely to affect the reputation of the company and the brand is likely to be rejected. Use of artificial Intelligence techniques can alert the producer regarding possible faults resulting in quality issues. Whether major or minor the faults will definitely influence the quality of the product, which in turn affect the manufacture and production and so identification of these faults at an early stage and their rectification highly essential.

Use of high resolution cameras in the AI solution of Machinery vision can identify the defects in a better way than to the naked eyes of human beings. With the support of an automatic response generating efficient Cloud-based data processing framework model, the performance of products can be monitored. Services required for the products when these enter the market can also are predicted. Strategic decisions required the survival of their products can also be formulated.

By using ML in semiconductor manufacturing McKinsey & Company have got a profit of 30% for the company by 30% [42]. This was possible by optimizing operations and reducing scrap rates. The manufacturing equipment used by the company incorporated with ML gives 10% cheaper annual maintenance costs, while reducing the downtime by 20% and the respective reducing inspection costs by 25%.

4.4 Inventory and Storage

The quality of inventory is essential to ensure successful customer service. Hence, inventory management plays an important role in achieving business' ambitions [33]. The annual cost of the storage of one unit of inventory can vary between 15 and 35% of its value.

Dealing with Inventory requires knowledge about managing stock volume, inventory space optimization, problem of accurate forecasting, managing stock volume, inventory space management, dealing with unnecessary and unused stock and customer satisfaction. To plan and maintain inventory at lowest cost along with customer satisfaction are required to have a successful supply chain. Managing inventory and avoiding risk of ineffectiveness in it requires the help of technology such as ML and AI which in turn will catalyze the business to survive effectively.

Stock tracking has been the prime aspect in inventory management. Use of ML in this process can improve the accuracy and transparency in the management process. With business specific parameters, predictions and calculations can be optimized for better results. The performed results can be used later for making smart decisions in future.

Use of ML in stock tracking and management increases business' focus on prime aspects like product quality and customer experiences. This results in improved business performance. Stock level is a major concern in inventory management. Inventories have definite chances of getting damaged or to be expired. Excess stock too results in waste of money. Uncertain stock levels need accurate predictions for upcoming demands and ML has the potential to take care of it.

In a storage unit with variation of inventories, sorting and storing needs to be idle in order to utilize the maximum space within in order to optimize the rental bills. Such technology can be easily provided using ML with image detection and additional customizations.

4.5 Transportation and Distribution

Technological developments have accelerated the transport sector's journey of innovation and evolution. AI is an example of an upcoming technology that has vastly contributed to the sector. This part of the supply chain is important as the inventory security depends on the transportation logistics. ML can be used to reduce traffic congestion and accidents, decrease carbon emissions, and also minimize the overall financial expenses.

Transportation routes in traditional SCM are established in advance and rarely changed. However, in the world with heavy competition in the business market where profit is a must, smart routing decision making needs to be applied. Use of ML powered transportation algorithms can predict optimal routes for transportation and plan appropriate resting stops for the journey using factors like weather, traffic, government regulations etc. Such algorithms can set threshold values for the journey using which a well-defined software can trigger alerts to the stakeholder in case real time value exceeds the given, this shall eliminate any further delays.

Ml algorithms can be used to ensure safer routes. In the cases, where the goods truck needs to travel within cities, burglary is a major concern. Significant steps like alerting the driver when entering such zones, planning of such routes in daylight etc. can be taken care of by the algorithms.

ML algorithms can be used to predict suitable vehicles for the entire route based on route constraints, goods to transport, minimum cost etc. Using this approach a cost efficient solution can be deduced for businesses that are yet to plan their means of transport.

Enhancing customer experiences should be a prime aspect of any business. Herewith, Estimated Time of Arrival (ETA) should be given the priority to ensure it. In the era where time is considered money, goods aren't the only concern for the clients. Getting efficiency in predicting ETAs is an ever learning task and ML algorithms are the premium solutions. Some parameters like river-route mapping, tonnage, pickup windows, delivery windows, no-entry time windows, unavoidable delays can be used for ETA prediction [22].

Use of ML in advance can be useful in determining the durability of the vehicles. Data necessary to check various parts of the vehicles and embedding them with ML algorithms can be used to predict vehicle breakdowns. A "work-in-progress" tech having potential to open new doors of possibility under transportation is SCM.

5 Deep Learning and Artificial Neural Networks in SCM

Machine learning (ML) is a subset of artificial intelligence which deals with self- improving computer algorithms. The models involved are based upon the training data set used and these models predict or make decisions without being pre-programmed.

As described by Arthur Samuel," it is a field of study that gives the ability to the computer for self-learns without being explicitly programmed". This means the machines gain knowledge without using any pre-developed code.

As stated above, neural networks are processing devices designed by the structure and functionalities of animal brain. These models are very fast, flexible and are suitable for real-time systems. Also, they support parallel architecture.

Deep Neural Networks (DNN) is different from other machine learning approaches from the perspective that they can process natural data in their raw form. Although deep learning concept was not getting good results for many years after its inception, two papers by G. Hinton [21, 24] acted like catalysts and since then it has flourished and there is no looking back. It develops its own feature vector whereas other approaches in ML use external experts with considerable domain expertise to derive such a feature vector which is used to represent the raw data internally. DNNs follow representational learning, which is a set of methods that allows a machine to be fed with raw data such that the representation is automatically discovered for detection or classification. For a wonderful presentation on the origin and developments on DNN the paper [27] by the three scientists LeCun, Bengio and Hinton has no contender. In fact, this paper published by them in the celebrated journal NATURE, perhaps was instrumental in their being awarded the coveted ACM Turing award in 2018.

In DNN, each method has multiple levels of representation and is obtained as a composition of simple but non-linear modules. Each individual module transforms from one level to a higher level which is more abstract than the previous one. The process of composition thus helps in learning complex functions in this way.

Using DNN algorithms several problems which could not be handled properly by using other AI techniques before its advancement have been solved. Because of its efficiency in handling high dimensional data it could be used in different domains of government, business and science. By the way it had outclassed other ML techniques in potential drug molecules prediction and particle accelerator data could be analyzed.

Several applications of deep neural networks (DNN) have been carried out in several application areas. One such area is the field of image processing [1]. DNNs are useful tools for object classification. One such classification direction is that of audio signals. An updated elaborative discussion on audio signals is presented in the literature [9]. One of the earliest applications of deep learning is the use of feature learning in image retrieval. An architecture has been proposed in this direction [19]. Sign language has its own applications for specific part of human society. But the curiosity arises as how to recognize these languages. The techniques proposed so far, is discussed at length in [37]. Retrieval of images is an important component of DNN and several such approaches exist in literature [19]. In addition, a text based approach is presented in [39]. Sentiment analysis is an important component of study of human behaviour. The use of RNN in sentiment analysis is also proposed [7]. The development of DNN has led to several advancements in the field of artificial intelligence (AI) [4]. A model called the S-LSTM-GAN, which is a shared recurrent neural networks are also proposed [2]. Generative adversarial networks (GAN) are advanced models of DNN and a single image super resolution technique using GAN is proposed [3]. A simplified presentation of Convolutional Neural Networks, its components and working principles and concepts associated is done in [29]. Age and gender estimation in images is a difficult but very useful task. A Wide ResNet-Based Approach is proposed in [16] to achieve this.

In the process of supply chain management the customers want to get what they want and if they don't get it are likely to move to the competitor. So, when multiple layers of analysis and computing power of DNN are put to work it leads to reduction of the mismatch between the products and demands of customers. This is in contrast to machine learning algorithms which are based upon single input and single outputs. DNNs have multiple levels of interactions between algorithms that take in and process varied data stream simultaneously.

Most accurate forecast is possible when algorithms parse buying habits of customers, current market trends, may be weather patterns also, leading to most accurate forecast possible. Multiple inputs can be handled by algorithms consistently leading to consistently fresh insights which allow adapting and pivoting on customers' demands by the way getting better of the competitors. Deep learning has adaptability as its strength and also takes care of pattern recognition.

Since ANNs work on the same model as the human brain, which has been studied for generations, the processes and inner workings can be broken down to a level that any audience can grasp to get the insights being used now to streamline their supply chain. Inventory management, logistics network routing, or demand planning have their future in deep learning.

In addition to the results coming out of an algorithm the user is now interested in knowing how that algorithm came up with the results it did. Seemingly inexplicable results are no more taken for granted. Since ANNs in general and DNNs in particular work on the same model as the human brain, the process of result generation can be broken into the steps of inner workings leading to the results and this is what very much essential in the context of supply chain in order to streamline them. Hence, in all the areas involved in supply chain like inventory management, logistics network routing, or demand planning, deep learning and neural networks are the future.

6 Use of Machine Learning Techniques in Amazon's SCM

Supply chain leaders can get ahead of supply chain disruption by adopting the innovative and agile practices of digital giants and startups [20]. Let's take Amazon, a multinational technology company which is primarily focused on e-commerce, cloud computing etc.

Amazon has changed traditional SCM and has set a leading example for using ML in SCM.

As stated by Statista in [14], As we can observe in Fig. 9 amazon had a net revenue of below $7 billion back in 2005, it has increased it to around $1.3 Trillion in a span of 15 years, which comes out to an increase of its revenue by over 185 times!! [38].

This was only possible due to its adoption of aggressive supply chain strategies along with its deployment of innovative technologies. Many policies such as Fulfillment by Amazon (FBA) and Fulfillment by Merchant (FBM) that came to effect also are a major reason. But, here the focus is on how they used ML in SCM to solve the problem of optimizing nearly every aspect of the supply chain—from inventory management and warehousing to delivery times and prices.

Amazon's Demand forecasting uses ML and AI heavily, which is termed as probability level demand forecasting, where vendors are made to take a decision which will impact their stocks directly. As stated by Jenny Freshwater "We're calling this before you click buy because we buy customer obsession in the weeks, days months before you're on website shopping and decide you want to buy something." Although it is very complex, Demand Forecasting is the root of Supply Chain and ML is heavily used in it. For example, from the forecasting data they can predict that in a particular

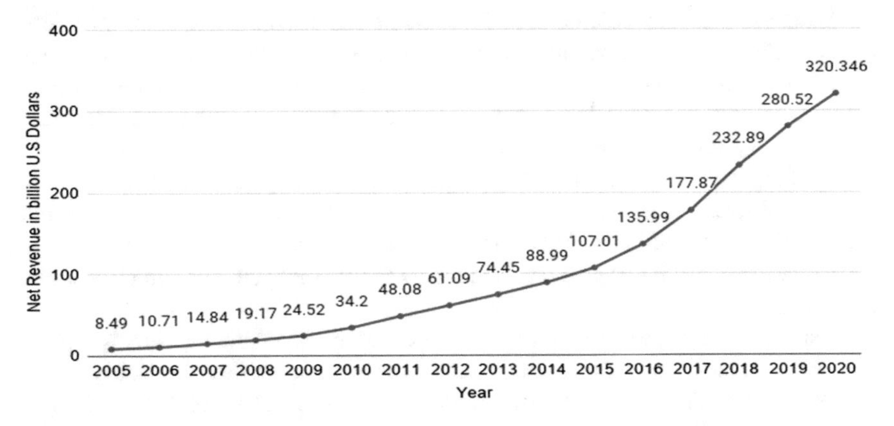

Fig. 9 History of net revenue of Amazon

region coffee beans will be in demand. As a result they load the warehouse well in advance with the required product so that there won't be a shortage and super-fast deliveries can be possible.

Another major factor where ML plays a role in supply chain is with the warehousing strategy. Not only do they outsource their inventory in a strategic way so that the same day deliveries are possible but also it is astonishing to know that over 50% of their products come from 3rd party sellers [26]. But a large part of Amazon's success is attributed to its excellent warehousing strategy, which ensures products are easily accessible almost anywhere on Earth.

Their warehouses are placed in locations which have more demand which are identified by ML models, and they also have mini-warehouses so as to meet the fast delivery needs.

Many of the users are loyal to amazon due to their crazy fast delivery, be it delivery on the same date, or in two days. They use numerous strategies which allow the company to get orders faster, easier and more efficiently to basically everywhere in the world—even remote and rural areas not served by traditional options. Differentiating between the types of delivery and prioritizing the order accordingly makes this possible.

One major Step that has a huge effect on amazon's revenue is their own manufacturing, Due to the enormous data that they have they are able to predict what a large chunk of customers want, and hence manufacturing the same, and who would buy a product from a different retailer when you have Amazon verified symbol on the product that you are looking for!

Their highly efficient ML and AI technology has enabled them to keep their overall per unit supply cost to a bare minimum—other company with far lower sales volumes and only their own warehouses are having trouble competing as a result.

As it has been observed above, there are many advantages on how ML is a boon to SCM and a company as a whole if used to the full extent. But nothing is perfect or foolproof with so many advantages, ML in SCM is bound to have drawbacks, here are few of them.

While the company uses demand forecasting they need to have more than 50% certainty that the product will have sales in a point of time [12]. And when hundreds of millions of products are dealt with it is bound to deal with equally enormous data. Amazon is dealing with a forecasting of approximately 140 data points for every product every day and every year [12].

Handling such huge data requires people with good expertise and shifting from old technologies to ML based ones takes a lot of time, effort and capital, although ML. Even though AI based SCM are much better it is not self-sufficient as it is necessary to control it and give required inputs for seamless outputs, which is not an easy task.

Systems become more and more complex day by day. Although this is a good sign, it might backfire as if an error occurs it will take a lot of time to identify where the issue lies. With such hi-tech systems it is necessary to make sure that the security aspects are equally good. Otherwise, the systems will be prone to attacks and the entire company might be shut down if something goes really bad.

7 Conclusions

In these tech savvy times, generation of data is an involuntary function. ML plays the role of an asset to any enterprise by processing this heap of valuable information. In this chapter we discussed various ML models and their use cases; we surveyed the evolution of SCM over a decade, proposed various possible ideas for integrating ML discreetly for each department of SCM and finally studied how Amazon applied this technology in its operations where its benefits and drawbacks were identified.

Research and advisory company Gartner [37] predicts that at least 50% of global companies in supply chain operations would be using AI and ML related transformational technologies by 2023. This showcases the growing popularity of ML in SCM. Businesses need to start reforming their strategies and start investing ML to their operations in order to gain the fruits of it.

8 Scope for Future Research

Technically speaking, profound study by [28] asserted suitable models under various scenarios wherein neural networks were appointed maximum number of times by various papers due to its ability to model prediction, clustering and classification problems. Use of DNN techniques in its various forms will definitely improve the results in SCM as mentioned by us in Sect. 6 above. But, very little research has been done and in fact no concrete work is found in the literature. This opens an entire horizon and in years to come we expect many fruitful applications of DNN techniques in SCM.

References

1. Adate, A., Tripathy, B.K.: Deep learning techniques for image processing. In: Machine Learning for Big Data Analysis, Berlin, Boston, De Gruyter, pp. 69–90 (2018)
2. Adate, A., Tripathy, B.K.: S-LSTM-GAN: Shared recurrent neural networks with adversarial training. In: Proceedings of the 2nd International Conference on Data Engineering and Communication Technology, Springer, Singapore, pp. 107–115 (2019)
3. Adate, A., Tripathy, B.K.: Understanding single image super resolution techniques with generative adversarial networks. In: Advances in Intelligent Systems and Computing, Springer, Singapore, vol. 816, pp. 833–840 (2019)
4. Adate, A., Tripathy, B.K., Arya, D., Shaha, A.: Impact of deep neural learning on artificial intelligence research. In: Deep Learning Research and Applications, De Gruyter Publications, pp. 69–84 (2020)
5. Altexsoft: Demand Forecasting Methods: Using Machine Learning and Predictive Analytics to See the Future of Sales (2019). https://www.altexsoft.com/blog/demand-forecasting-methods-using-machine-learning/
6. Anderson, M.G., Katz, P.B.: Strategic sourcing. Int. J. Logist. Manag. **9**(1), 1–13 (1998)

7. Baktha, K., Tripathy, B.K.: Investigation of recurrent neural networks in the field of sentiment analysis. In: Proceedings of IEEE International Conference on Communication and Signal Processing, pp. 2047–2050 (2017)
8. Ballou, R.H.: The evolution and future of logistics and supply chain management. Eur. Bus. Rev. **19**(4), 332–348 (2007)
9. Bose, A., Tripathy, B.K.: Deep learning for audio signal classification. In: Deep Learning Research and Applications, De Gruyter Publications, pp. 105–136 (2020)
10. Bousqaoui, H., Achchab, S., Tikito K.: Machine learning applications in supply chains: An emphasis on neural network applications. In: 2017 3rd International Conference of Cloud Computing Technologies and Applications (CloudTech), pp. 1–7 (2017)
11. Chuprina, R.: AI and machine learning in manufacturing: the complete guide (2020). https://spd.group/machine-learning/ai-and-ml-in-manufacturing-industry
12. Clark, G.: Inside Amazon's AI-powered supply chain, retailbiz (2019). https://www.retailbiz.com.au/latest-news/inside-amazons-ai-powered-supply-chain/
13. Columbus, L.: How machine learning is redefining supply chain management (2019). https://erpblog.iqms.com/how-machine-learning-is-redefining-supply-chain-management
14. Croom, S., Romano, P., Giannakis, M.: Supply chain management: an analytical framework for critical literature review. Eur. J. Purchas. Supply Manage. **6**(1), 67–83 (2000)
15. Daugherty, P.J.: Review of logistics and supply chain relationship literature and suggested research agenda. Int. J. Phys. Distrib. Logist. Manag. **41**(1), 16–31 (2011)
16. Debgupta, R., Chaudhuri, B.B., Tripathy, B.K.: A wide ResNet-based approach for age and gender estimation in face images. In: Proceedings of International Conference on Innovative Computing and Communications, Springer, Singapore, pp. 517–530 (2020)
17. Devi, G.N.R., Solanki, V.K., Makkar S.: Applications of machine learning techniques in supply chain optimization. In: ICICCT 2019—System Reliability, Quality Control, Safety, Maintenance and Management, pp. 861–869 (2019)
18. Farmer, D.: Purchasing myopia—revisited. Eur. J. Purchas. Supply Manage. **3**(1), 1–8 (1997)
19. Garg, N., Nikhitha, P., Tripathy, B.K.: Image retrieval using latent feature learning by deep architecture. In: Proceedings of IEEE International Conference on Computational Intelligence and Computing Research, pp. 1–4 (2014)
20. Hippold, S.: Supply chain leaders can get ahead of supply chain disruption by adopting the innovative and agile practices of digital giants and startups (2021), 4 Ways to Compete against Nontraditional Supply Chains (gartner.com)
21. Hinton, G.E., Osindero, S., Teh, Y.-W.: A fast learning algorithm for deep belief nets. Neural Comput. **18**, 1527–1554 (2006)
22. https://www.getfareye.com/insights/blog/transportation-2020-5-ways-ml-will-drive-higher-levels-of-operational-efficiencies(transportation)
23. Jiangtao, R., Sau, D.L., Xianlu, C., Ben, K., Reynold, C.D.C.: Ninth IEEE International Conference on Data Mining, Naive Bayes Classification of Uncertain Data (2009)
24. Johnson, T.: How the Amazon supply chain strategy works (2020). https://tinuiti.com/blog/amazon/amazon-supply-chain/
25. Koh, L., Demirbag, S.C., Bayraktar, M., Tatoglu, E., Zaim, S.: The impact of supply chain management practices on performance of SMEs. Ind. Manage. Data Syst. **107**(1):103–124 (2007)
26. Leblenc, R.: How Amazon is changing supply chain management (2020). https://www.thebalancesmb.com/how-amazon-is-changing-supply-chain-management-4155324
27. LeCun, Y., Bengio, Y., Hinton, G.: Deep learning. Nature **521**, 436–444 (2015)
28. Lummus, R.R., Vokurka, R.J., Alber, K.L.: Strategic supply chain planning. Prod. Invent. Manag. J. **39**(3), 49–58 (1998)
29. Maheshwari, K., Shaha, A., Arya, D., Rajasekaran, R., Tripathy, B.K.: Convolutional neural networks: a bottom-up approach. In: Deep Learning Research and Applications, De Gruyter Publications, pp. 21–50 (2019)
30. Marinagi, C., Trivellas, P., Sakas, D.P.: The impact of information technology on the development of supply chain competitive advantage. Proc. Soc. Behav. Sci. **147**, 586–591 (2014)

31. Markets and Markets: Supply Chain Analytics Market by Component, Software (Supplier Performance Analytics, Demand Analysis and Forecasting, and Inventory Analytics), Services, Deployment Model, Organization Size, Industry Vertical, and Region—Global Forecast to 2025 (2019). https://www.marketsandmarkets.com/Market-Reports/supply-chain-analytic-market-139106911.html
32. Markets and Markets: Smart Manufacturing Market by Enabling Technology (Condition Monitoring, Artificial Intelligence, IIoT, Digital Twin, Industrial 3D Printing), Information Technology (WMS, MES, PAM, HMI), Industry, and Region—Global Forecast to 2025 (2019). https://www.marketsandmarkets.com/Market-Reports/smart-manufacturing-market-105448439
33. Min, H.: Artificial intelligence in supply chain management: theory and applications. Int. J. Logist. Res. Appl. **13**(1), 13–39 (2010)
34. Morgan, J., Monczka, R.M.: Supplier integration: a new level of supply chain management. Purchasing **120**(1), 110–113 (1996)
35. Movahedi, B., Lavassani, K., Kumar, V.: Transition to B2B e-marketplace enabled supply chain: readiness assessment and success factors. Int. J. Technol. Knowl. Soc. **5**(3), 75–88 (2009)
36. Panetta, K.: Gartner Predicts 2019 for Supply Chain Operations (2018). https://www.gartner.com/smarterwithgartner/gartner-predicts-2019-for-supply-chain-operations
37. Prakash, V., Tripathy, B.K.: Recent advancements in automatic sign language recognition (SLR). In: Computational Intelligence for Human Action Recognition, CRC Press, pp. 1–24 (2020)
38. Sabanoglu, T.: Annual net sales of Amazon 2004–2019 (2020). https://www.statista.com/statistics/266282/annual-net-revenue-of-amazoncom/
39. Singhania, U., Tripathy, B.K.: Text-based image retrieval using deep learning. In: Encyclopedia of Information Science and Technology, 5th edn, IGI Global, USA, pp. 87–97 (2020)
40. Tan, K.C.: A framework of supply chain management literature. Eur. J. Purchas. Supply Manage. **7**(1), 39–48 (2001)
41. Tank, T.P., Crum, M., Arango, M.: Benefits of inter firm coordination in food industry supply chains. J. Bus. Logist. **20**(2), 21–41 (1999)
42. Utermohlen, K.: How machine learning (ML) is transforming manufacturing, towards datascience (2018). https://towardsdatascience.com/how-machine-learning-ml-is-transforming-manufacturing
43. Wenzel, H., Smit, D., Sardesai, S.: A literature review on machine learning in supply chain management, Chapters from the Proceedings of the Hamburg International Conference of Logistics (HICL). In: Kersten, W., Blecker, T., Ringle, C.M. (eds.) Artificial Intelligence and Digital Transformation in Supply Chain Management: Innovative Approaches for, Artificial Intelligence and Digital Transformation in Supply Chain Management: Innovative Approaches for Supply Chains, Proceedings of the Hamburg Int. vol. 27, pp. 413–441 (2019)

Role of Big Data in Supply Chain Management

V. Srividya and B. K. Tripathy

Abstract In today's world, the supply chain management sector plays a vital role in everyone's life. During the recent times, the number of people or customers buying or ordering goods online has increased enormously. The management process of transformation of raw materials to finished goods can be termed as Supply Chain Management (SCM). The actors involved in the supply chain are the vendors or suppliers, distributors and customers. At every stage of this chain, large volumes of data get generated. These data are a collection of information from a variety of domains such as goods, clothing, accessories and so on. This big data need be used wisely to improve the supply chain management. The big data is a more than just internal data from Enterprise Resource Planning (ERP) and SCM. The statistical analysis methods such as regression, hypothesis testing or sample size determination are used to analyse the internal as well and newly created data that provide new outcomes which in turn help to improve the decision making involved in the supply chain. Decision making choices might be which operating model to choose, who should be the vendor for a particular item and so on. This chapter aims to explain the role of Big data and its analysis in supply chain management.

Keywords Enterprise resource planning (ERP) · Supply chain management (SCM) · Big data · Statistics · Big data analytics

1 Introduction

Integration of key business process with the supply chain is recognized as Supply chain management [6]. The management of the supply chain can be done in a smooth manner if there is a common set of rules in the supply chain process across the

V. Srividya
School of Computer Science and Engineering, VIT, Vellore 632014, Tamil Nadu, India

B. K. Tripathy (✉)
School of Information Technology and Engineering, VIT, Vellore 632014, Tamil Nadu, India
e-mail: tripathybk@vit.ac.in

K. Perumal et al. (eds.), *Innovative Supply Chain Management via Digitalization and Artificial Intelligence*, Studies in Systems, Decision and Control 424,
https://doi.org/10.1007/978-981-19-0240-6_3

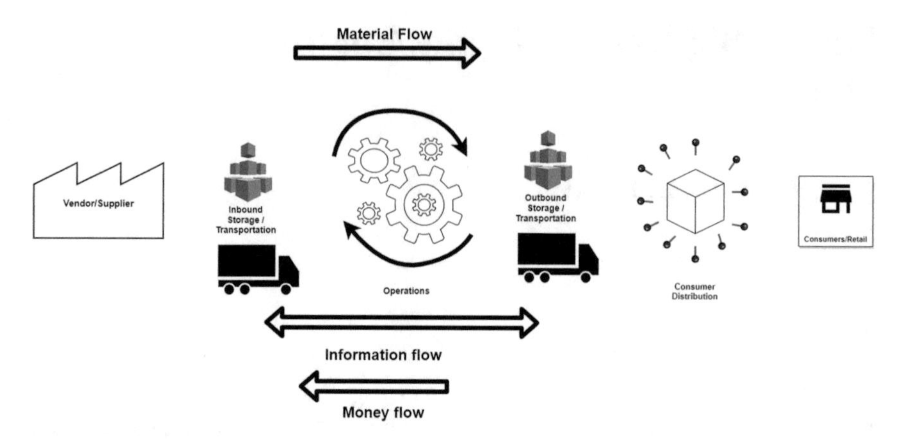

Fig. 1 Supply chain management process

key businesses. The process starts with the vendors or suppliers and end with the customers. Figure 1 shows the supply chain management process.

In this figure, three flows are shown; namely the material flow from the vendors or suppliers to the customers or retail. Money flow from the customers to the suppliers and information flow is bidirectional. In other words, the information flows to and from the vendors or suppliers.

The process workflow starts with the vendors or suppliers dealing with the raw materials. A local storage or inbound storage is a place within the supplier's premises to keep the chosen raw materials, ready to transport to the manufacturing plant. In the manufacturing plant, all the production operations occur to create the finished product. These finished products are placed in a warehouse or an outbound storage. From these warehouses the finished goods are transported to the Point of sale locations or retail stores or consumer distribution centres. From the distribution centres, the goods are shipped to the customers or end user might buy them from the retail stores.

2 Eight Key Processes of Global Supply Chain

The eight key processes that comprise of the global supply chain management [3] are, Customer relationship management, Customer service management, Demand management, Order fulfilment, Manufacturing flow management, Procurement, Product development and Commercialization and Returns as shown in Fig. 2.

The customer relationship management deals with the identifying the target customers and the customer groups based on the goal of the business firm. Special offers or discounts or customized agreements are created with these regular customers in order to maintain a healthy relationship with the customers. Profit from each

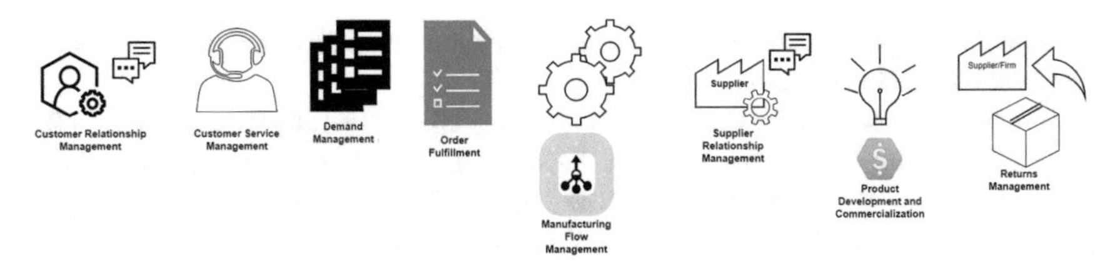

Fig. 2 Key processes of global supply chain

customer is monitored and also the impact of such value-added services to the customers on the company's finance is closely observed.

The customer service management is the interface that appears to the customer. This interface provides information such as order status, product availability, shipping details and so on. All this information is real-time. The administration of the Products and Services Agreement (PSA) is the responsibility of the customer service management.

The Demand management's task is to maintain the balance between the rate of demand of products and the rate of supply from the company. A mechanism to forecast the demand for a particular product has to be developed in order to prevent any inconvenience to the customer due to the lack of the product availability. The demand management team must work closely with the production, procurement and distribution team.

The order fulfilment involves a smooth integration of the production, logistics and the marketing departments. The aim of the order fulfilment is to satisfy the user's order. While maintaining the order fulfilment, the cost involved in the delivery of the product must be kept minimum or optimal.

The manufacturing flow management involves handling the production efficiently, making the production runs flexible, determining the boundaries of operation, identifying any manufacturing constraints or requirement constraints.

Supplier relationship management is process of the company managing the relationship with their suppliers for certain type of products. Each relationship with a supplier is associated with a PSA.

Product development and commercialization deals with making new products quickly and efficiently. The time to market of the product must be kept at the minimum in order to make the firm a big success.

Returns management is an important feature every firm must contain. In order to get better with time, the returns must also be managed such that the loss incurred reduces over a period of time by providing products that have a very low chance of failure in the market.

Another major domain that is being used in this chapter is the field of Big data and the Analytics associated with it.

Big data can be defined as data but with large volume whose growth is exponential in nature. Such data cannot be efficiently stored or processed using the traditional data handling techniques. In order to gain insights or learn patterns from such large data, systematic computational analysis-based tools were developed that use statistics to discover the hidden patterns.

Many of the techniques for standard data sets cannot be directly extended to the stage of big data in obvious manner. As a result, separate algorithms have been proposed to handle the situations. For the sake of illustration, we present here some examples. Big data clustering and classification algorithms are presented in [11]. Uncertainty has been an integral part of modern-day datasets. There are several models like fuzzy set, intuitionistic fuzzy sets, rough sets and neighbourhood rough sets to deal with uncertainty in datasets. Fuzzy set-based clustering algorithms for big data are discussed in [9, 13]. Similar algorithms for intuitionistic fuzzy techniques are discussed in [10]. Rough set and neighbourhood systems-based clustering algorithms for big data are discussed in [12]. Neighborhood based knowledge acquisition using Map Reduce from big data over Cloud Computing is discussed in [8]. An elaborative article in [14] discusses with several data clustering for big data. Image processing can be handled as an application of clustering. Also, many times images provide big data sets as inputs. Deep neural network techniques used for image segmentation are presented in [1]. As the Euclidean distance used as similarity measure is applicable to datasets which are linearly separable. But, for non-separable data sets a different type of measure called kernels are used. Algorithms using hybrid models obtained through suitable combinations of several models have been found to be more efficient than the algorithms developed by using the individual components. So, in an attempt to develop clustering algorithms for big data by using hybrid uncertainty-based models, kernels as similarity measures and following the possibilistic approach is proposed and experimentally verified to be more efficient than many clustering algorithms developed on big data in [7]. The algorithm proposed is a robust one. Big data visualization is an important aspect and has many challenges before it. A comparative analysis and presentation of the challenges faced there in have been presented in [5]. Social networks are in elaborate use in society in now a day. Also, the data involved in the communication among members of these social networks is huge and complex. So, data mining techniques were to be modified and improved to fill in the requirements. A discussion on such techniques is presented in [15].

3 Characteristics of Big Data

The 4V's forming the main characteristics of Big data are namely: Volume, Variety, Velocity, Veracity in Fig. 3 [2].

3.1 Volume

One of the primary characteristics associated Big Data is volume. In today's world data generated ranges from terabytes to zettabytes. Main challenge faced by the IT

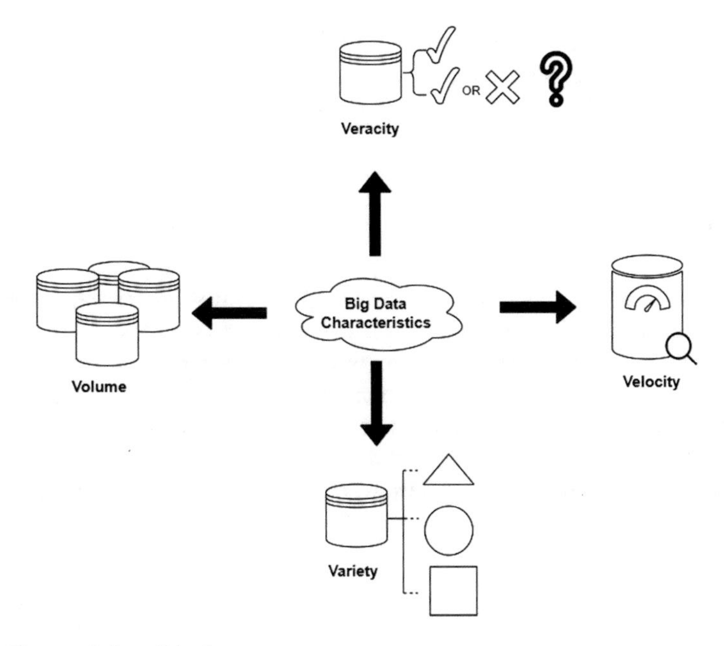

Fig. 3 Characteristics of big data

firms is their capacity to efficiently process the large volumes of data. Using Big data analytics one can smoothly process the data and generate insights.

3.2 Velocity

Velocity can be defined as the rate at which the data is generated and in turn the rate of processing, storing and analyzing the data. Companies need to possess infrastructure that support such speeds. Some examples could be uploading 900 images in a day on a platform or collecting streams of tweets for a certain topic.

3.3 Variety

Another major characteristic of Big data is variety. Data is not always structured in nature. There are mainly three main types of data namely the structured data, semi-structured data and unstructured data. Structured data can be described as the data that can be stored in the form of rows and columns such as a relational database. SQL can be used to query these types of data. The semi-structured data is a data that cannot be organized into rows and columns but do possess some order. Aim of the unstructured data is to ease space. Example of semi-structured data is XML files.

Un-structured data cannot be organized in any order and is not associated with any existing models. Business Intelligence systems and analytics use such unstructured data to gain insights. Examples of un-structured data are videos, audio, word, pdf.

3.4 Veracity

The data collected is not always clean, there might be unwanted or unrelated data that have to cleaned. The analysis is performed on the clean data to prevent any bias in the accuracy due to noise. Consider a stream of twitter data, the line extracted contains lots of characters such as punctuation marks, extra symbols such as "@@@", "###" and so on. While performing text analytics the first step is to clean data by removing stop words such as is, are, not and punctuation marks as well. Hence veracity of data is a very critical characteristic that must be observed before analyzing data.

4 Big Data Analytics

Big data analytics are broadly divided into two categories based on the goal namely Decision making and action-oriented. Decision making based analytics deal with business intelligence and generating insights that help in making better business-related decisions). Action-oriented analytics deal with quick analysis to recognize patterns in data based on which some actions need to be triggered [16]. Figure 4 shows some techniques of big data analytics depending on the input data.

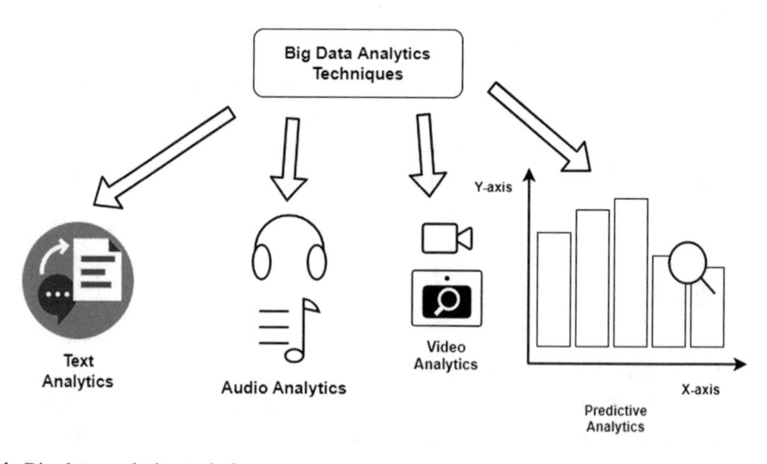

Fig. 4 Big data analytics techniques

4.1 Text Analytics

Text Analytics is also known as text mining and can be defined as a technique that involves analysing unstructured text in order to extract related information. Examples of textual data include social media data, emails, chats, online forms, company logs. Models or techniques used in text analytics are statistical methods, machine learning and computational linguistics. The results from the analysis help in decision making. The decisions may be recommending a product to a user based on their review of a similar product.

Some of the text analytics methods include Information extraction (IE), Relation Extraction (RE).

4.2 Audio Analytics

Speech Analytics can be defined as techniques that take in un-structured data such as audio or speech files as input and extract valid information from them. The systems where audio analytics is useful would be healthcare centres, customer service call centres, customer interaction with devices such as Alexa, Siri and so on. The recorded calls from the customer service centre are analysed to understand the mood of the user while they were on call, the behaviour of the user can also be determined using the tone in which they communicate. Information such as product complaint, service issues are also learnt from the calls. These calls are also used as a way to measure the performance of the customer service agents and reward them accordingly. There are two major methods in Speech analytics namely the Transcript-based approach and the Phonetic-based approach. The transcript-based method works by transcription of the audio file using automatic speech recognition techniques. The phonetics methods use phonemes as a way to differentiate words. For example, cat and bat can be distinguished based on the pronunciation.

4.3 Video Analytics

Video content analysis (VCA) is a way to extract significant information, analyse and monitor the video files. Major systems that make use of VCA is the Police department that rely data from the closed-circuit television (CCTV). The videos recorded by the CCTV help the police to understand the crime scene well.

CCTV is also placed at corners or near shelves in the store. Details such as time spent by the user at the certain shelves can be used to keep similar items in locations nearby. Stalls may be re-arranged to make it comfortable for the user to shop. Video analytics techniques are of three types namely server-based architecture, edge-based architecture and Agent Vi's distribution architecture. In server-based architecture, the camera-based videos are stored in a backend server where they would be processed. In edge-based, the analysis is performed at local hard disks. Agent Vi's architecture is a mixture of both server based and edge based.

4.4 Predictive Analytics

Analytics that help in predicting an outcome that might happen in future based on insights from historical data. Some of the well-known predictive analytic techniques include Neural Networks (NN), Support Vector Machines (SVMs), Decision Trees, Linear Regression, Clustering, Logistic Regression, Association Rules. Predictive analytics can be applied in any domain provided the amount of historical data for that domain is available in large amounts. For example, the future demand for a certain product such as stationary or some biscuits or some clothes apparel. Predictive analytics is classified into two streams namely Regression techniques and machine learning techniques.

5 Role of Big Data in Supply Chain Management Design

Figure 5 shows the design of Supply chain management. In the next few paragraphs, these areas are explained in detail [17].

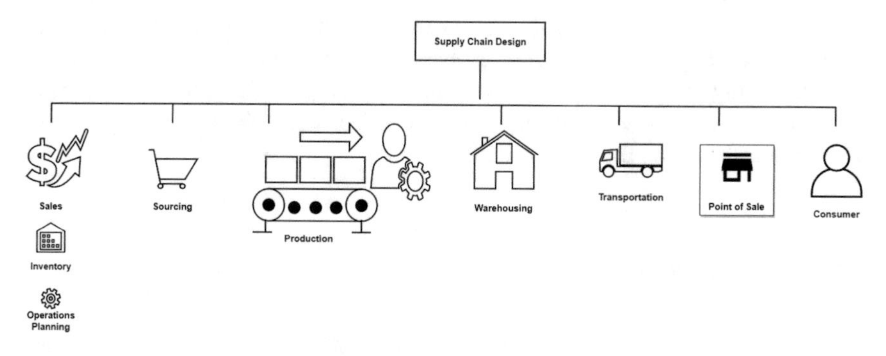

Fig. 5 Design of supply chain management

5.1 *Sales, Inventory and Operation Planning Area*

Planning phase of the supply chain can be termed as the utmost data-driven process. The inputs to the planning process include data from Enterprise Resource Planning (ERP) and Supply Chain Management (SCM) planning tools.

In order to enable real-time demand and supply balance, the planning phase must be redefined. The inputs to this mechanism could also be taken from external sources in addition to the ERP and SCM tool inputs.

Point of sale (POS) data, production volume, and inventory data associated with planning can be analysed using statistical methods. The aim of this analysis is to be able to identify any imbalance between demand and supply. The results from the study may help make better decisions such as changing the price, adding new features, modifying the promotion time and so on.

Some companies such as Blue Yonder, IBM and Amazon have used forecasting methods to predict the rate of demand on their products. This enables them to plan their rate of supply in advance. The frequency of stockouts and lack of proper inventory levels reduces. This in turn increases the company's profit.

One of methods to shape demand also includes recommending similar products to users that are in stock thereby satisfying the customer's order.

5.2 *The Sourcing Phase*

In the sourcing process, the data gathered include supplier information and procurement volumes. The data associated with the supply process does not confine to the traditional spend analysis or the performance review of the annual supplier. There must be a way to identify any deviations from the normal pattern of delivery. A predictive risk management technique can help make better strategic decisions if the firm gets information about the trend in the market, strike rate in social media or any information on the financial status of their suppliers. For example, if the firm gets information about their supplier's chance of getting bankrupt in the near future, alternate plans can be kept ready in advance thereby reducing the chances of any loss in future.

One approach to streamline the process would be to create a database comprising of factor costs, commodity prices and plant utilization information. A predictive model using this database helps in choosing the right suppliers to package their products based on certain criteria.

Another database corresponding to the costs involved in warehousing and transportation can be created. The contents of this database are the calculated costs of Logistics Solution Providers (LSPs) and warehouse rents across the globe. This data can be used while negotiating with the companies to make the best deal.

5.3 Manufacturing Process

The manufacturing process includes scheduling production runs, monitoring quality of products or defects in products. Big data analytics techniques can be used to plan the production run schedules beforehand by considering any fluctuations in electricity pricing or power failures or lack of raw material. An analysis can be done on the data of defected products to understand the basic cause for the defect in order to prevent them in future.

The quality of products in the agricultural industry can be analysed using sensors. The result of analysis is then used to maintain or improve the product quality.

In case machines used in the manufacturing process have some issue, the sensor associated with it will immediately trigger an alarm to the authorities enabling them to kept a spare ready in case the machine stops working in future.

5.4 Warehousing Process

Warehouses are places where finished goods are stored. The choice of the right warehouse such that the utilization of space in the warehouse is efficient, the transportation to destination is minimum or an automatic rack bay to arrange the goods according to the schedule for the next day is a very critical decision to make. Big data analytics can be used to help the firm choose the right option for the warehouse such that it is cost effective and satisfies the other criteria. Data of staff at the warehouse that manages and also picks the products to ship them to their destinations is stored. The analysis on the data helps in future staff recruitment or shift staff between warehouses to balance the work load efficiently.

Every warehouse will have transportation that shift stocks between warehouses or ship them to their destination. Using the latest technologies such as sensors on these trucks, real-time information can be collected such as products in the truck, destination of the truck, number of times the product has been shipped to either the destination or a warehouse. Using this information one can increase the stock of that product in the source warehouse or if the destination is another warehouse, the stock in the second ware house can be increased thereby reducing the number of trips from the source warehouse. This in turn reduces the transportation cost and speeds up the process of delivery at the second warehouse. This real-time data can be merged with the ERP data and Warehouse Management System (WMS) to identify any additional waste in the warehouse process.

5.5 *Transportation Phase*

In the transportation phase, the trucks are loaded with the products from the warehouses or from vendors/ suppliers or deliver products to destination. The factors involved in transportation are the amount of fuel consumption, waiting times to collect goods from warehouse, waiting time while delivering the products to the customer [4].

The real-time data from the sensors placed in the truck collects the amount of fuel used by the vehicle per day, per week, per month or per year. These sensor data are analysed and a calculated estimate of fuel consumption is derived from the results. The cost for fuel is calculated using the estimate of fuel consumption in order to have an estimated budget for transportation.

The GPS on the trucks are synced with the bays of the destination warehouse to keep the stock ready for pickup to reduce the waiting time.

If certain products are delivered to the same destination from a long time, this information can be used to identify the probable time at which the customers will be at home to collect the product, thereby reducing the number of re schedules of the delivery and hence reducing the transportation cost involved in delivery re-schedules.

There amount of carbon emitted during every drive can be reduced to a great extent if all the travels are scheduled in advance and also prioritized. A limit can be set on the number of travels a particular truck makes per day. This also helps maintain the vehicles in the long run.

5.6 *Point of Sale (POS)*

Point of Sale can be retail stores or supermarkets where the finished goods are placed for the customers and go and buy. The decision involved here includes which products to keep at their shelves, which products to keep at the aisle, pricing of the products at their store and so on.

Based on the data comprising of previous transactions at the shop, an analysis is performed to understand what kind of products get out of stock frequently. This implies that those products have high demand in that locality and increasing the stock of those goods in the store would increase the profitability.

Decisions such as combo offers for products based on buying pattern of the customers, price reduction of products that are relatively less in demand, duration of keeping the items in store, choosing the right non-perishable products and so on are very critical. Methods as Market basket analysis can be used to make the right decisions.

For large stores, drone cameras can be used to monitor the inventory levels on the shelves. Camera outside the store monitors the cars in the parking to anticipate the number of customers who might come to the store. Based on these real time data, the necessary arrangements are made in the store.

A major challenge in these stores is to detect out-of-stock products and also prevent them in order to reduce the inconvenience caused to the customers. One solution includes an alarm trigger at the billing counter if a particular item that is sold frequently does not show up for a long time. This trigger ensures a person to check the item in the shelf. In case it is not in stock an immediate re-stock action is ordered to ensure smooth functioning of the store.

Another solution could be to have light weight sensors on the shelves that monitor the inventory stock in the shelf. In case there is a shortage of any product, a trigger should enable automatic re-stock process of the product in demand.

6 Big Data Analytics Approach in Digital Supply Chain Management

6.1 Big Data Analytics in Team Creation

The first step in setting up Big Data Analytics (BDA) as a part of the existing supply chain in business as shown in the Fig. 6 is to establish a team with good big data analysts. The skills that a member of this team must possess are a blend of extremely deep analytical skills and the ability to work with unstructured data along with a detailed understanding of the supply chain of the business [18]. These data scientist's skill set also includes ability to develop solutions, integrate these solutions into the system, have knowledge on mathematical and statistical methods. This team is the main contributor in finding solutions to supply chain problems that involve big data by rapidly developing prototypes. The role of the BDA team involves creation of plans for each idea that is generated in the next stage to be tested and deployed. The team must schedule regular meetings with the business groups who are directly involved in designing the solution. Clear deadlines for each idea must be set in order to streamline the prototyping process. Training sessions must be held in the firm in enable the employees to be comfortable while sharing their problems with the data

Fig. 6 Algorithmic steps in supply chain management

Algorithm: Big Data Analytics in Supply Chain Management

1: Create a special team for Big data Analytics.

2: Identify the Supply chain problems in business.

3: Map the data sources and identify external use cases.

4: Start setting the priorities in business.

5: Build Solution.

6: Evaluate Solution.

7: Repeat steps 2 to 6 quarterly.

Table 1 Phase-wise contributors list

Phase-wise contributors list			
Serial No.	Phase	Contributors	Time duration
1	Big data analytics team creation	• Head of supply chain • Supply chain HR • CIO (to review)	Approx. 2–3
2	Identify supply chain problems	• Head of supply chain • Supply chain leads/department • Supply chain leadership chain	Approx. 2 weeks Note: 2 and 3 are done in parallel
3	Map data source and find external use cases	• Head of supply chain • Supply chain leads/department • Supply chain leadership team • Data scientists	Approx. 2 weeks Note: 2 and 3 are done in parallel
4	Setting business priorities	• Head of supply chain • Supply chain leads/department	Approx. 2 weeks
5	Build solution	• Big data analytics team • Supply chain leads/department (to review)	Weeks to months
6	Evaluate solution	• Supply chain leadership team • Big data analytics team • CIO	Approx. 1 week (quarterly review)

scientists. The contributors involved in this phase are the Head of the supply chain management team, the HR lead of the supply chain stream who selects the right candidates for the role and the Chief Information Officer (CIO) who will review the shortlisted candidates. The time it takes for to create the entire team is approximately 2–3 months as shown in Table 1.

6.2 *Identification of Supply Chain Problems in Business*

The second step is to identify the problems in the supply chain as shown in Fig. 6. In order to generate profitable ideas, the BDA team must understand the current business priorities and the performance gaps. With the help of the present supply chain management that has been already implemented in the organization, a good understanding of the pros and cons of system would already be known. This information can be used the BDA team start thinking of ideas that would improve the supply chain. Some of the areas of concerns would be any forecasting issues with respect to delivery or demand or out-of-stocks for some specific retailer. Product reviews can also make of the BDA techniques to improve the product of future releases. The people involved in identifying the problems are the Head of the supply chain management team, every lead of the supply chain and the leadership team of the supply chain. This step is combined with the next step and the total time taken is approximately 2 weeks as shown in Fig. 6.

6.3 Mapping of Data Sources and Finding External Use Cases

The third step is to map the data sources and find out the type of data (structured or unstructured). It also includes identifying external use cases. The starting point to come up with possible solutions would be to look out for already existing re-usable solutions. These ideas could be sourced from the fellow competitors, or professional organizations or software retailers. An internal team can also be setup who would provide suggestions in choosing the right processes that might help in improving the supply chain using big data. Ideas could be collecting more and more data to make the forecasting more reliable and accurate or real-time testing results can be used as a feedback to improve the existing solution and so on. Once the team starts listing out the possible ideas, the company must find prospective data sources that might help to facilitate the idea. These data sources could be existing data from ERP or SCM or data that can be got from external sources such as trading partners, government agencies or Google. Data can also be generated using sensors installed at shelves in the retailer's store or at a section of the manufacturing unit. A proper review of the potential sources of data must be done to ensure all useful sources are taken into consideration. Some ideas may be very expensive to implement but risk must be taken as without testing, the idea cannot be rejected. Table 2 shows a list of important data sources with respect to SCM applications for most companies.

6.4 Setting Priorities in Business

The fourth step in the process is to set up the priorities in the business as shown in Fig. 6. Once the list of valuable ideas, possible data sources is ready, an estimate for each idea's solution comprising of the cost, implementation feasibility, and potential gain from that solution must be prepared. A high level draft of the each solution is created. The contents of these draft are the process to access the data source, the methods used to analyse the data and usage of the observations obtained from the analysis. An estimate of the impact is made by an expert of that domain. While prototyping the ideas, a clear criteria showing the service improvement level, cost estimate and the capital involved that support the solution must be defined. This helps

Table 2 Important data resources in SCM

Data source	Type of data	Size of data	Structure type	Quality of data	Owner-ship
ERP systems	Demand, sales Capacity SC plans	Large	Structured	Medium	Private
Barcode/RFID scan	Location Time ID	Large	Structured	High	Private
Sensors/Cameras	Quality, image Humidity Temperature Status of parts	Large	Structured and unstructured	High	Private
Archives	Financial statistics Price data weather data	Medium	Structured	Medium/High	Public and private
Internet	Hits, clicks Streams Statistics Comments	Very large	Unstructured	Low/Medium	Public and private
Social networking	Preferences Text Developments	Very large	Unstructured	Low	Public and private

in segregating work that needs to be continued and the ones that must be discontinued. A priority list of ideas is created by comparing each idea's cost, difficulty level, degree of impact and strategic importance. BDA team works on these plans every quarter.

6.5 Build Solutions

The fifth step in the process is to start building the solutions based on the priority list created in the last step. The solution phase must regularly involve the end-user and get feedback while developing the solution to ensure that the solution addresses the underlying problems which might have been left out otherwise. "Hackathons" are a way to quickly get initial prototypes ready in hours. As the solutions get fine-tuned by the feedback from the users, the solution becomes more reliable and can be confidently released to the end users once completed.

6.6 Evaluate Solution

Once the solution is ready, the impact of the solution must be evaluated to understand the impact of the product. The performance can be compared against the targets that lead to the birth of the solution. Criteria for evaluation include business impact as well as tested hypotheses. The stakeholders decide on the next step based on the evaluation results. The next steps may be expansion of the idea, full implementation of the idea or rolling out the final product. Evaluation may happen in iterations depending on some idea's complexity level. Finally, based on the review comments modifications are made to the supply chain priorities that would be handled in the next iteration of development.

7 Conclusions

In this chapter, an overview of the supply chain design, big data, characteristics of big data, big data analytical techniques along with role of big data in supply chain design and the approach to start using big data as a part of the supply chain are explained. By using big data analytics in SCM, the gains of the firm can be improved to a great extent as the company would be able to predict their demands in advance and also the number of complaints with respect to product quality or out-of-stock issues, packaging issues can be reduced to a great extent. Overall cost in developing products would be optimized. Companies can stop making certain products based on the feedback from end users which otherwise the company might not be aware of and spend too much on it.

References

1. Adate, A., Tripathy, B.K.: Deep learning techniques for image processing. In: Bhattacharyya, S., Bhaumik, H., Mukherjee, A., De, S. (eds.) Machine Learning for Big Data Analysis, pp. 69–90. De Gruyter, Berlin, Boston (2018)
2. Anuradha, J.: A brief introduction on Big Data 5Vs characteristics and Hadoop technology. Proc. Comput. Sci. **48**, 319–324 (2015)
3. Ballou, R.H.: The evolution and future of logistics and supply chain management. Eur. Bus. Rev. **19**(4), 332–348 (2007)
4. Croom, S., Romano, P., Giannakis, M.: Supply chain management: an analytical framework for critical literature review. Eur. J. Purchas. Supply Manage. **6**(1), 67–83 (2000)
5. Divya, Z.G., Tripathy, B.K.: Comparative analysis of tools for big data visualization and challenges. In: Anouncia, S., Gohel, H., Vairamuthu, S. (eds.) Data Visualization, Springer, Singapore, pp. 33–52 (2020)
6. Croxton, K.L., García-Dastugue, S.J., Lambert, D.M., Rogers, D.S.: The supply chain management processes. Int. J. Logist. Manag. **12**(2), 13–36 (2001)
7. Tripathy, B.K., Mittal, D.: Hadoop based uncertain possibilistic kernelized c-means algorithms for image segmentation and a comparative analysis. Appl. Soft Comput. **46**, 886–923 (2016)

8. Tripathy, B.K., Vishwakarma, H.R., Kothari, D.P.: Neighbourhood based knowledge acquisition using MapReduce from big data over cloud computing. In: Proceedings CSIBIG14, pp. 183–188 (2014)
9. Tripathy, B.K., Deepthi, P.H.: Handling Fuzziness in Big Data using Clustering Techniques, NCICT-15, Bangalore (2015)
10. Tripathy, B.K., Deepthi, P.H., Mittal, D.: Hadoop with intuitionistic fuzzy C-means for clustering in big data. Adv. Intell. Syst. Comput. **438**, 599–610 (2016)
11. Seetha, H., Tripathy, B.K., Murthy, M.K. Modern Technologies for Big Data Classification and Clustering, IGI Edited volume (2017)
12. Tripathy, B.K.: Rough set and neighbourhood systems in big data analysis. In: Sugumaran, V., Arun Kumar, S., Arun Kumar, T. (eds.) Computational Intelligence Applications in Business Intelligence and Big Data Analytics. CRC Press, Taylor & Francis Group, Chapter-10 (2017)
13. Tripathy, B.K., Deepthi, P.H.: An investigation of fuzzy techniques in clustering of big data. In: Sugumaran, V., Arun Kumar, S., Arun Kumar, T. (Eds.), Computational Intelligence Applications in Business Intelligence and Big Data Analytics. CRC Press, Taylor & Francis Group, Chapter 11 (2017)
14. Tripathy, B.K., Seetha, H., Murthy, M.K.: Uncertainty based clustering algorithms for large data sets. In: Modern Technologies for Big Data Classification and Clustering, IGI Edited volume, Chapter 1, pp. 1–33 (2017)
15. Tripathy, B.K., Sooraj, T.R., Mohanty, R.K.: Data mining techniques in big data for social network. In: Panda, M., Hassanien, A.E., Abraham, A. (eds.) Edited volume, Big Data Analytics: A Social Network Approach, p. 21. Taylor and Francis Publisher, Chapter-3 (2018)
16. Vashisht, P., Gupta, V.: Big data analytics techniques: a survey. In: 2015 International Conference on Green Computing and Internet of Things (ICGCIoT). IEEE, pp. 264–269 (2015)
17. McKinsey & Company: Big data and the supply chain: the big-supply-chain analytics landscape (Part 1) [Online] (2021). https://www.mckinsey.com/business-functions/operations/our-insights/big-data-and-the-supply-chain-the-big-supply-chain-analytics-landscape-part-1
18. McKinsey & Company: Big data and the supply chain: the big-supply-chain analytics landscape (Part 2) [Online] (2021). https://www.mckinsey.com/business-functions/operations/our-insights/big-data-and-the-supply-chain-the-big-supply-chain-analytics-landscape-part-1

Classification and Analysis of Market Segment in Supply Chain System for Stationary Products

P. K. Yeshwanth Kumar and Kumaresan Perumal

Abstract A leading stationary company one of the key market player among the stationary goods manufacturing companies has a wide range of products. The company being an international market player has a hefty market share of 40% globally. In India the company's products have not obtained such a big market share when compared to potential competitors. There are various reasons for the company's lower market share. The key aspect is to classify, characterize and analyze their product mix at a region level in order to anticipate the most profitable product segment which can aid in devising new marketing strategy and better segmentation.

Keywords Product mix · Sales versus quantity · Product placement · Regional analysis

1 Introduction

The world's leading Stationary Company manufactures wide range of products for writing, drawing and creative design, and decorative cosmetics. They are the global enterprise in the fields of learning and playing, Marketing, General writing, Graphic collections and Premium Art. The firm's mission statement "A companion for life" emphasizes the firm's enthusiasm towards it products which are designed for consumer benefit and the improvement of quality of life [1]. Perceived as an innovative pacesetter and an inventor organization. The research is carried out as regional studies in Kanchipuram and Vellore district in Tamil Nadu for product sales analysis, product quantity analysis and product placement analysis. The outcome is the product mix analysis to fetch least moving product with least profit.

P. K. Yeshwanth Kumar
Moorea Commerce Fruits, Rungis, France

K. Perumal (✉)
Vellore Institute of Technology (VIT), Vellore, Tamil Nadu, India
e-mail: pkumaresan@vit.ac.in

K. Perumal et al. (eds.), *Innovative Supply Chain Management via Digitalization and Artificial Intelligence*, Studies in Systems, Decision and Control 424,
https://doi.org/10.1007/978-981-19-0240-6_4

2 Literature Review

The stationary products market in India has valued around INR 40 billion during 2013–2014, witnessing a growth of around 17.5% per year during 2010–2014. The market leader of the paper stationary is ITC with brands Classmate, PaperKraft followed by Navneet around 7%. ITC has classified the market segment as PaperKraft for corporate and Classmate for children and students. The writing industry is growing at a rate of 8–10% p.a. Of that the top 3 market players in the pen category is in such a way that Cello Pens at the top followed by GM Pens with its brands Reynolds putting the Linc Pens & Plastics Ltd to third position. The other market players are Faber Castell, Koukyo Camlin, Flair, Montex, Lexi, Mont Blanc, Parker Pens etc. In general out of all stationary product brands the key player—all-rounders' are Faber Castell, Koukyo Camlin and Staedtler [2].

Stationary Products are commodity products. Price becomes the major differentiator as an attribute in determining consumers' buying decision. There is an obvious low brand loyalty condition because the consumers' priorities are (1) Characteristic features like color and purpose of the product, (2) Price of the product, and the 3rd and 4th positions to Brand and the attractive looks respectively. Retail outlets play a major role in brand building because if a particular brand's product doesn't occupy a better shelf space then consumers' will go for brand alternatives. Also in these products consumer involvement is poor because majority of the consumers' doesn't communicate the brands they utilise indicating the need for better Below the Line (BTL) promotions. Impulse purchase category is at the maximum level in these product categories [3].

The contribution of the organised and unorganised sectors is 75 and 25% of total market value. Among stationary products pens market share has the highest. In that pens with cost of Rs. 15 and less contribute 60% share, pens with a cost range of Rs. 15–100 has 25% share and Premium pens contribute to 15% of market share. Market segmentation is the key indicator of product growth rate among competitors. Analysing India's literacy rate before market segmentation is the most important criteria [4].

The general segmentation includes (1) Scholastic (primary and secondary school going), (2) Adolescent (graduate, post graduate and professional courses), (3) Frequent users (in office, commercial and public establishments), (4) Occasional users (like a multitude of housewives and literate manual workers), (5) Professionals. There have been a limited number of studies conducted for this market [5].

The technological advancements and improved telecommunication equipment have adversely affected the need for writing instruments. As a consequence there should be a phenomenal effect due to the increasing literacy level and the economic activity edifying the demand fetching a limited and moderate level of raise in this market.

3 Product Classification

The products were classified into six main categories for the purpose of analysis. They are art products, craft products, general writing, fine writing, technical writing and stationary. The products that are involved in the studies are wax crayons, erasable crayons, oil pastels, sketch pens, poster colors, fun paints, color pencils, paint brushes, art cart kit, pencils, mechanical pencils, leads, scales, geometry box, erasers, glue sticks, child safe scissors, sharpeners, stencils, canvas kit, modeling dough, 3 dimensional outliners, polychromatic artist color pencils, pitt monochrome graphite and charcoal, art grip aquarelle, water colors, artist gift box, perfect pencils, ball pens, text liners, multi markers, stamp pad and ink. All products used in the studies are domestic range of products [6].

4 Regional Studies

In this study totally 2 districts were covered involving 10 regions—towns. Region 1 is Tamil Nadu, Vellore district with 4 prime towns namely Walajapet and Ranipet, Arcot, Sholingar, Sathuvachari and Katpadi. Region 2 is Kanchipuram with 5 towns namely Kalpakkam, Sriperumbudur, Chengelpet, Chinnakanchipuram North Kanchipuram. The purpose of choosing these regions for the study are (1) The number of shop dealers and the stationary outlets in these locations are high. (2) To analyze the purchasing power and sales aspect in mediocre districts. (3) To make a comparative analysis of two different regions—one region with maximum number of educational institutes being proportional to increased sales ratio and one region with comparatively less educational institutes with higher shop dealers.

5 Product Sales Analysis

The process of analysis involves segregation of products according to the type of classification as mentioned with the specificity of region. The tool used for segregation and analysis is Microsoft Excel [7]. Raw data of the real time dealer orders and sales were obtained from the regional sales manager of a period of 2 months—March and April. The purpose of these 2 month data accusation is (1) suitable time period for peak sales as these are exam months for schools and colleges. (2) Cumulative data of yearend sales of the prime locations can be a potential source (Tables 1, 2 and Figs. 1, 2).

Table 1 Vellore district analysis data

Category	Quantity in no	Sales in rupees
Art	205	14,635
Craft	243	12,237
Fine writing	840	2931
General writing	496	8148
Stationary	1492	24,701
Technical	2446	40,732

Table 2 Kanchipuram district analysis data

Category	Quantity in no	Sales in rupees
Art	940	4239
Craft	394	20,279
Fine writing	789	3518
General writing	789	15,475
Stationary	1656	17,110
Technical	3567	55,728

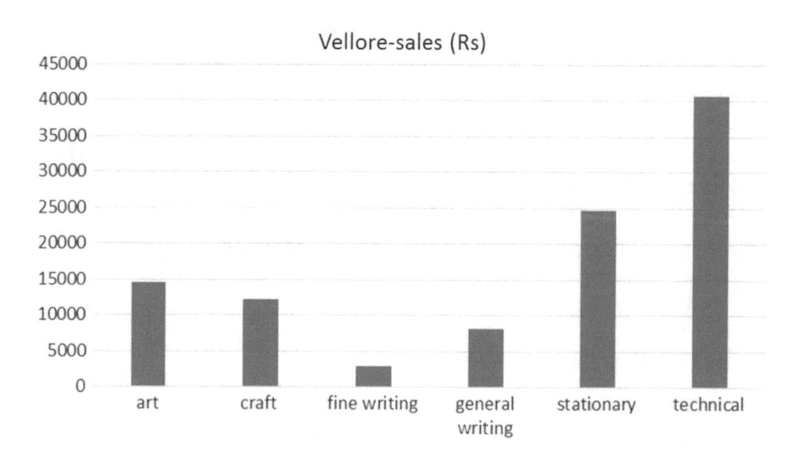

Fig. 1 Sales indicator

6 Product Quantity Analysis

The process of quantity analysis involved the same steps followed for sales analysis. Utilizing the same data the quantity of products sold in both regions were studied. The raw data had a collection of sales and quantity purchased from each and every shop dealer in both the districts. The purpose of quantity analysis is to identify which category products have the highest market share among its product mix in those regions [8] (Figs. 3 and 4).

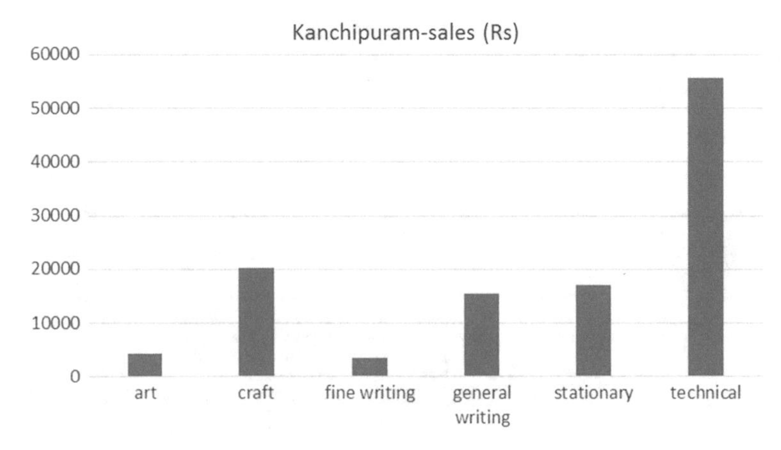

Fig. 2 Sales indicator

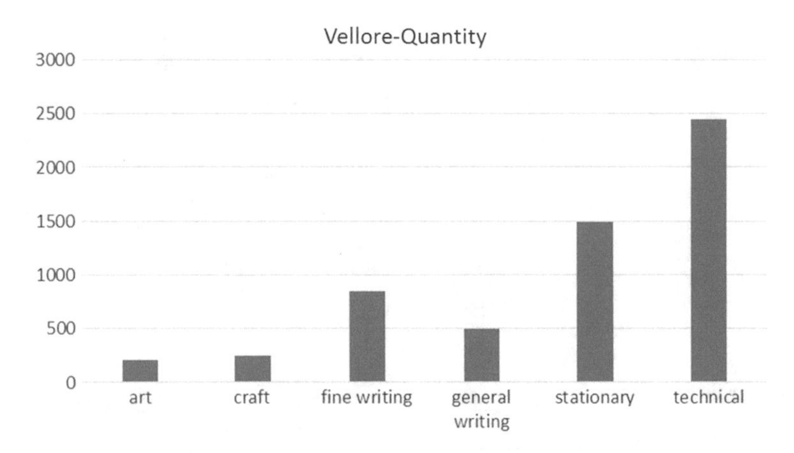

Fig. 3 Vellore district quantity indicator

7 Comparative Analysis

The process of studying the sales and quantity of the 2 different regions is to anticipate the product type that has been classified providing highest profit for the company and to identify the product that has been purchased in huge amount indicating the demands in these regions (Tables 3, 4 and Figs. 5, 6).

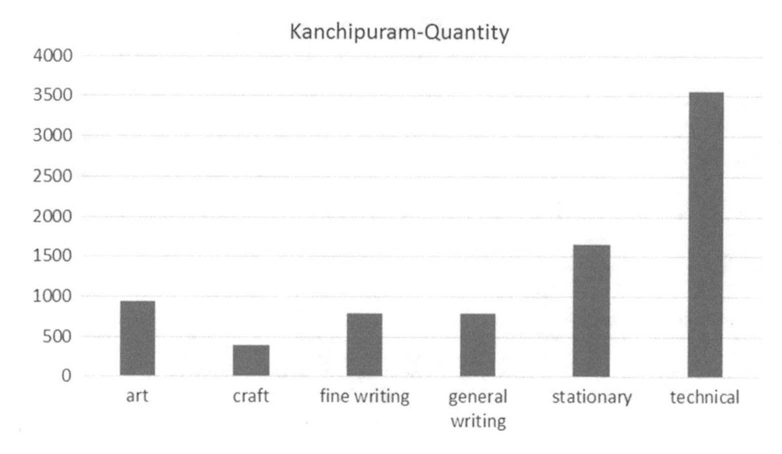

Fig. 4 Kanchipuram district quantity indicator

Table 3 Quantity indicator

Category	Vellore Quantity in no	Kanchipuram quantity in no
Art	205	940
Craft	243	394
Fine writing	840	789
General writing	496	789
Stationary	1492	1656
Technical	2446	3567

Table 4 Sales indicator

Category	Vellore Sales in rupees	Kanchipuram Sales in rupees
Art	14,635	4239
Craft	12,237	20,279
Fine writing	2931	3518
General writing	8148	15,475
Stationary	24,701	17,110
Technical	40,732	55,728

8 Product Placement Analysis

In the districts of Vellore and Kanchipuram each region included in the study along with the products were analyzed for their of level of placement in order to identify the product category that contributes the maximum sales and profit as well (Tables 5, 6 and Figs. 7, 8).

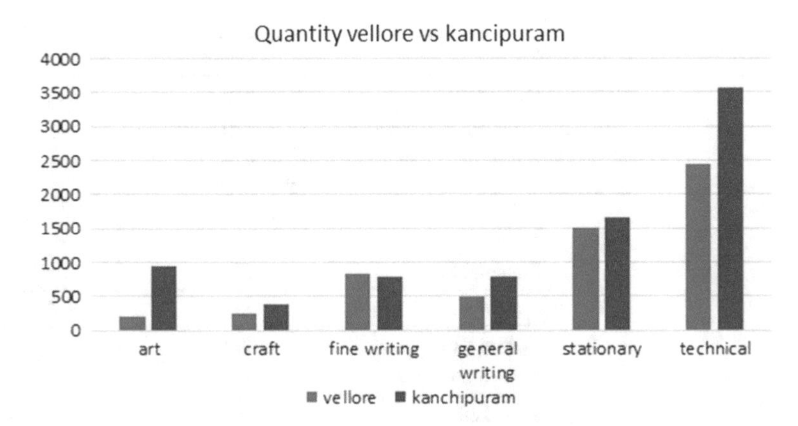

Fig. 5 Quantity indicator

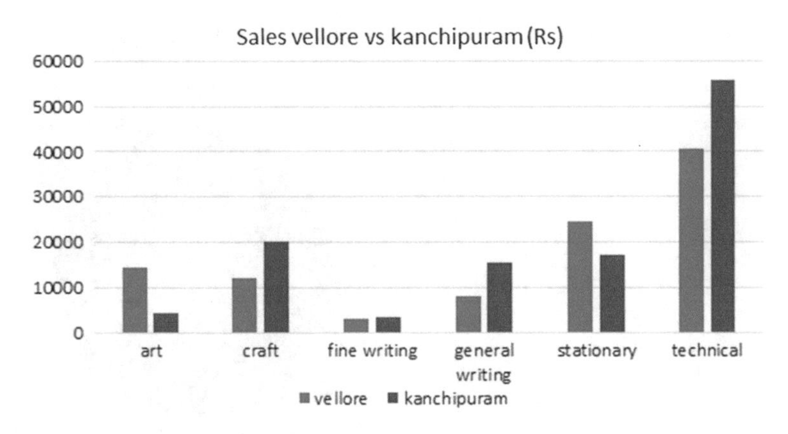

Fig. 6 Sales indicator

Table 5 Stationary products

Region	Quantity	Sales
Vellore	348	6214
Sathuvachari–Katpadi	452	6467
Arcot	315	5744
Walajapet	175	3348
Sholingar	202	2928
Kanchipuram 1	346	3694
Kanchipuram 2	1004	7335
Chengelpet	196	4232
Sriperumbhudur	12	255
Kalpakam	98	1594

Table 6 Technical products

Region	Quantity	Sales
Vellore	310	4707
Sathuvachari–Katpadi	840	14,696
Arcot	512	8665
Walajapet	200	2604
Sholingar	784	12,664
Kanchipuram 1	1972	28,210
Kanchipuram 2	239	4067
Chengelpet	800	12,423
Sriperumbhudur	200	3089
Kalpakam	356	7939

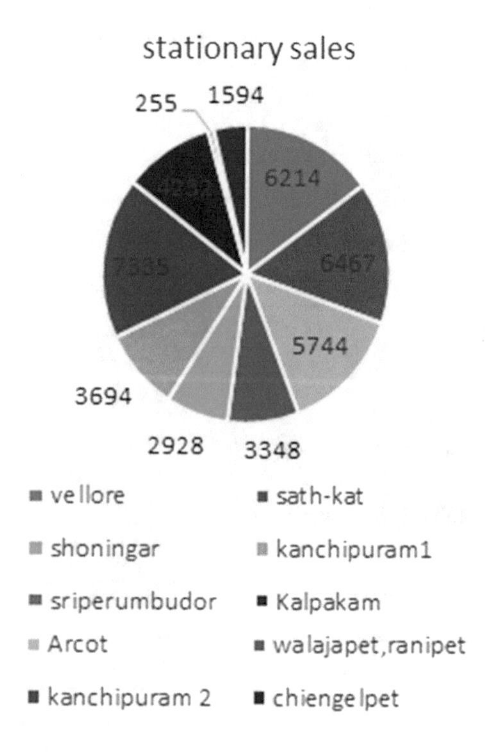

Fig. 7 Stationary indicator

9 Product Mix Analysis Outcome

The process of analysis clearly indicates that among various product categories stationary and technical contributes the highest as well placed product class. This provides a platform for the company to identify and strengthen their product mix. The studies have provided the least moving product category in terms of quantity

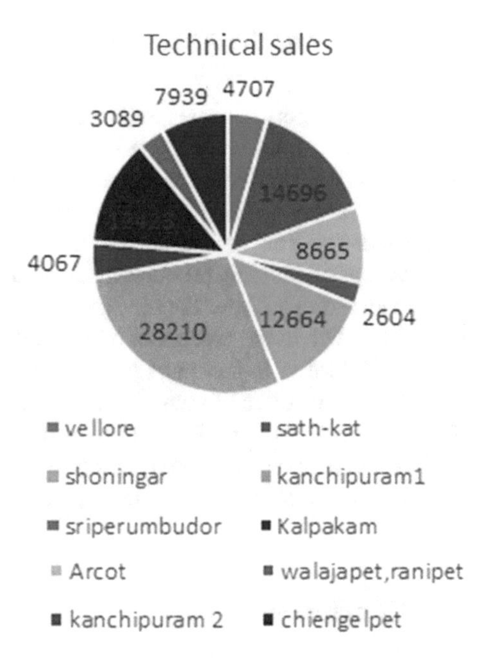

Fig. 8 Technical indicator

fetching least profit. It is the art and craft category that contributes for least sales indicating that it is the least placed product in these regions. The lower rate of product placement can be due to varied reasons out of that one major reason could be lack of products in art and craft product line. Developing new product line with innovative ideas or adding up new products to the existing product line can alleviate the product placement [8].

10 Limitations

The paper involves study on stationary products only without involving competitors' products. It covers only the domestic products and their sales excluding the international products. The products subjected in the paper have limited categorization and segregation suitable for limited data obtained for studies [9]. The regional analysis performed in the studies covers only 2 districts indicating the need for more area coverage. The drawback lies in the quantity of data obtained for the study. The quantity of products sold and the type of products sold in this particular region gives an analysis report specific to two regions, which cannot be the only source utilized for devising new marketing strategy lacking a holistic approach. Similarly the regional study analysed unique market segmentation which are specific to those two regions only which changes from region to region giving a limited analysis report.

10.1 Scope of the Future Research

The analysis report indicates the need for strengthening their product mix in order to yield a grip hold on market. It provides the platform for identification of the consumers' need in the regionally segmented market. There can be strategies devised to edify the segmentation process based on their product line. The placement mix can be enhanced in order to attain better consumer involvement. The study indicates the minimum profit yielding products in the product line where there is a need of product placement through Above the Line (ATL) and Below the Line (BTL) placement tools to attain growth in the process of profit yield. It projects the prediction of stationary products future in Indian market especially in the south with its product varieties [10]. The product line of art and craft segment is the iconic symbol of the company picturizing the position of the brand in Indian market. It requires re positioning of the brand in a perceptually mapped consumers' mind by expanding the product mix with more product class in writing segment [11].

11 Conclusion

The outcome of the analysis shows that art and craft category contributes least sales in the selected Vellore and Kanchipuram districts. It indicates the right product placement should be improved to increase the sales for the least moving category. At the same pace it is identified that art and craft can be a potential source of profit if targeted rightly. Both the regions included in the study have almost the same percentage of GDP. Kanchipuram has the highest profit in terms of sales while Vellore has the highest scope for better product placement.

References

1. Tsai, F.M., Bui, T.D., Tseng, M.L., Ali, M.H., Lim, M.K., Chiu, A.S.: Sustainable supply chain management trends in world regions: a data-driven analysis. Res. Conserv. Recycl. **167**, 105421 (2021)
2. Attaran, M.: Digital technology enablers and their implications for supply chain management. In: Supply Chain Forum: An International Journal, vol. 21, no. 3, pp. 158–172. Taylor & Francis (2020, July)
3. Abolghasemi, M., Beh, E., Tarr, G., Gerlach, R.: Demand forecasting in supply chain: The impact of demand volatility in the presence of promotion. Comput. Ind. Eng. **142**, 106380 (2020)
4. Singh, A., Jenamani, M., Thakkar, J.J., Rana, N.P.: Propagation of online consumer perceived negativity: quantifying the effect of supply chain underperformance on passenger car sales. J. Bus. Res. **132**, 102–114 (2021)
5. Witczyńska, K.: The impact of the electronic commerce market in the supply chain during COVID-19 pandemic in Poland. Eur. Res. Stud. **23**(2), 648–658 (2020)

6. Wang, C.H., Chen, J.Y.: Demand forecasting and financial estimation considering the interactive dynamics of semiconductor supply-chain companies. Comput. Ind. Eng. **138**, 106104 (2019)
7. Dutta, P., Suryawanshi, P., Gujarathi, P., Dutta, A.: Managing risk for e-commerce supply chains: an empirical study. IFAC-PapersOnLine **52**(13), 349–354 (2019)
8. Garay-Rondero, C.L., Martinez-Flores, J.L., Smith, N.R., Morales, S.O.C., Aldrette-Malacara, A.: Digital supply chain model in Industry 4.0. J. Manuf. Technol. Manage. (2019)
9. Seth, D., Pandey, M.K.: A multiple-item inventory model for a non-stationary demand. Prod. Plan. Control **20**(3), 242–253 (2009)
10. Mentzer, J.T., DeWitt, W., Keebler, J.S., Min, S., Nix, N.W., Smith, C.D., Zacharia, Z.G.: Defining supply chain management. J. Bus. Logist. **22**(2), 1–25 (2001)
11. Attaran, M.: Critical success factors and challenges of implementing RFID in supply chain management. J. Supply Chain Oper. Manage. **10**(1), 144–167 (2012)

Agile Supply Chain: Framework for Digitization

Chiranji Lal Chowdhary

Abstract The supply chain is a new business model for supply chain planning or transportation management that aims to enhance supply chain flexibility. Rather of having the resources and skills in-house, supply chain can be purchased as a service and paid for on a per-use basis. Service providers' specialization and focus enable them to achieve economies of scale as well as economies of scope, as well as lucrative outsourcing options. Industry 4.0 causes upheaval and forces businesses to reconsider how they build their supply chains. A number of new technologies have developed that are disrupting traditional working practices. Agility is a critical component of business success in complex industrial environments because it allows companies to compete effectively in a competitive market. Firms are increasing relying on information systems in achieving such agility throughout the supply chain. The chain becomes a fully connected ecosystem that is fully transparent to all parties involved, from raw material, component, and part suppliers through transporters of those supplies and completed items, and lastly to the customers expecting fulfilment, thanks to digitization. This chapter aims to explain the framework of digitization in agile supply chain management.

Keywords Agile · Supply Chain Management (SCM) · Lean · Digitization · Industry 4.0 · Internet of Things

1 Introduction

1.1 Agile

Agile project managements are iterative approaches and they deliver a project across the life cycle. Such agile life cycles consist of multiple iterations or gradual steps towards a project's completion. Iterative methods being used to encourage velocity

C. L. Chowdhary (✉)
School of Information Technology and Engineering, Vellore Institute of Technology, Vellore, India
e-mail: Chiranji.lal@vit.ac.in

K. Perumal et al. (eds.), *Innovative Supply Chain Management via Digitalization and Artificial Intelligence*, Studies in Systems, Decision and Control 424,
https://doi.org/10.1007/978-981-19-0240-6_5

and adaptability in software development projects which this lead to go along rather than following a linear path. The main objectives of an agile or iterative strategy are to gain benefits in the process by central values and behaviors of trust, flexibility, empowerment and collaboration. In software development, agile means discovering the requirements and developing the solutions with a collaborative effort from a cross functional and self-organizing teams, stakeholders and users [1–5]. Following are the main agile values:

- Individuals and interactions over processes and tools
- Working software over comprehensive documentation
- Customer collaboration over contract negotiation
- Responding to change over following a plan.

There is a possible categorization between agile and non-agile values in Table 1.

1.2 Supply Chain

The supply change management (SCM) can be understood as a strategy to manage the flow of goods and services which involves the movement, raw-materials storage, work-in-process inventory. This approach finishes the goods along with end to end order accomplishment from the point of source to the point of ingesting. Therefore, a supply chain management arrangements with the management of complete production flow of a service or goods, which initiates form raw components to the end product delivery to the consumers [6–13].

Currently, the supply chain is mounting rapidly everyday with augmented complexity. There is involvement of various factors to affect the functioning of a supply chain and they are listed below:

- Digitization,
- Change and its fundamentals,
- Augmented reality,
- AI and machine learning.

Lean supply chain is another influencing approach which is known for its effectiveness [14–17]. It is ill-fated results that many of the supply chain entities fail to

Table 1 Agile and non-agile values

Non-agile value	Agile value
Processes and tools	Individuals and interactions
Comprehensive documentation	Working software
Contract negotiation	Customer collaboration
Following a plan	Responding to change

understand and hold the agility concept. Basically, it relates to the use of competency, flexibility, speed and responsiveness to handle the activities of supply chain organizations very well eventually.

1.3 Agile Supply Chain

The solution to the many challenges existing in today's supply chain management networks is the agile supply chain which is a need for the company's success. This idea has been recognized as a solution in an evolving world to improve the responsiveness of a supply chain. The supply chain of nowadays is competing with different techniques, but agile approach is the most widely used. In a dynamic market demand trend, the agile supply chain approach anticipates demand fluctuations. This solution has made it easier for suppliers to manage their own data through self-service features [18–22].

In the present period, the agile supply chain approach has become the focused and mostly chosen strategy. The main concentrations are on the control of inventories. These techniques can avoid surplus inventory and future shortages. Nowadays when many software projects are launched in the industry for proper management of inventory, agile supply chains came as the best by getting more accurate and up-to-date demand. This is how easily the response time reacts to consumer demand and meets the end customer. In order to make it possible to have a network based on retailers and all other drivers of the chain linked electronically, this is how end-to-end visibility is accomplished. To make it a stable chain, it will be necessary to use state-of-the-art, planning applications that enable decision-making by working teams, and often with what if scenario capability, businesses typically go for contingency plans that are deemed extra costly than such applications make it more responsive and cost-effective.

The rest of this paper is organized as follows. Section 2 briefly overviews the involvement of agile in SCM and a proper flow diagram of agile supply chain management. Section 3 elaborates the agile benefits of Supply Chain in detail. The experimental results and the analysis are reported in Sect. 4. Section 5 is differences between Agile SCM, Conventional SCM and Lean Concepts. Digital Impacts of agile manufacturing process with AI tools or IoT integration are discussed in Sect. 6. Section 7 elucidates Case Study on Demand Variance on Responsive to market and Consumer demands with predictive analysis or quantitative analysis. In the end, Sect. 7 concludes this work and gives future research direction.

2 Agile in SCM with Proper Flow Diagram

Figure 1 expresses the flow of goods and services in Agile SCM process. Initially, all the goods from the suppliers arrive at a central store and stored in the warehouse of

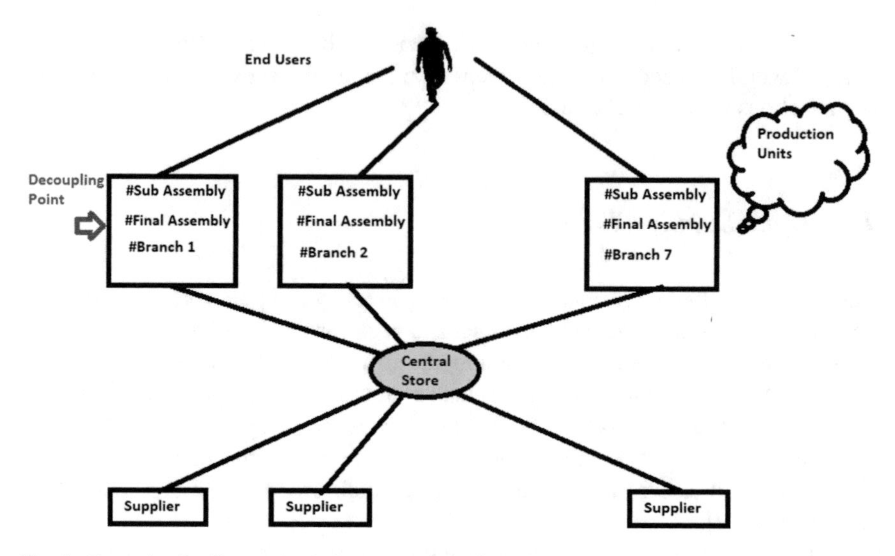

Fig. 1 Example of agile supply chain management flow of goods and services

the central store. The warehouse of the central store sends the goods to respective sub-assembly points and decoupled. The decoupled points are sent to the final assemblies. Then combination of sub-assembly and final assembly points are made as production point and sent to their respective branches. When goods reached the branches then they are delivered to the end-users. Overall this flow of the goods and services is an example of an agile supply chain management.

An agile supply chain is categorized in four main components as virtual, process alignment, network based and market sensitive. Figure 2 is a mind map of an agile

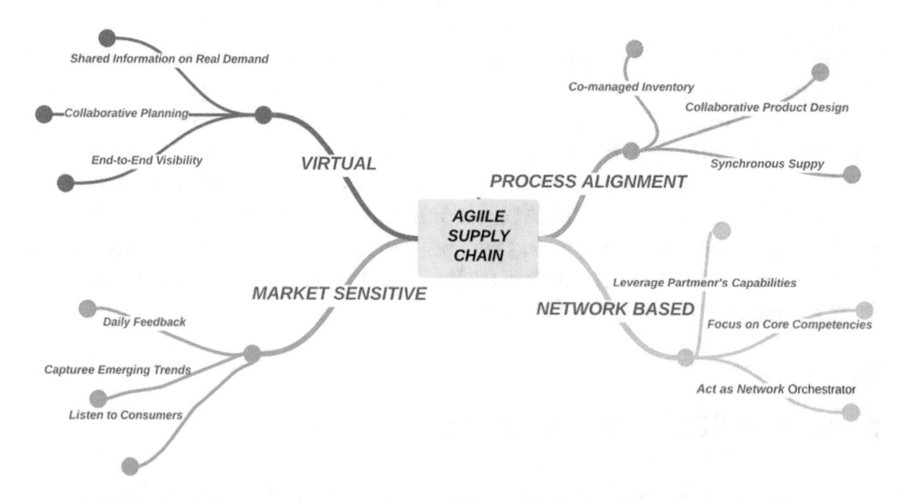

Fig. 2 Agile SCM

supply chain management. The virtual information is shared based on real demand, collaborative planning is done and end-to-end visibility is there. The process alignment inventory is co-managed, collaborative product design is done and supply is synchronized. The network based: partner's capabilities are leveraged; focus is mainly on the core competencies and acts as network orchestrator. Last, the market sensitive feedback is giving emerging trends which captures and listen to the customer.

3 Benefits of Agile Supply Chain

The switching attention from controlling costs to rising customer satisfaction has increasingly complicated the modern supply chain. Digitization, virtual reality, IoT, and artificial intelligence are now changing the manner in which a supply chain works and businesses that are unable to keep up with the pace are losing exciting deals. As an outcome, enterprises consider an agile supply chain that is extremely versatile and capable of adapting rapidly to evolving circumstances. Various benefits of agile supply chain are listed as:

- Network-wide Collaboration and Analytics: Shift focus to a left-to-right, pull-oriented supply chain driven by actual customer demand.
- Align and Balance Supply with Demand: Place high emphasis on incorporating demand, supply, logistics, and product design network partners in all aspects of network-wide collaboration.
- Sense and Respond: Invest in the ability to more quickly sense unexpected events, with the ability to globally adapt and respond to them.
- Ability to respond to the change effectively: This helps the respective companies to handle the volatile nature of the market, frequent changes in economy, changes and advancements made in science, equipment and respective technologies and other such features in a much better way. The companies can adjust to the volatile nature of the market conditions by helping the manufacturers in producing the exact correct amount of goods on a daily basis instead of traditional methods like relying on the monthly and quarterly forecasts.
- Mass customizations: Customers prefer customization as it gives them a sense of personal products and their possessions. For example, printing the customer's name on the coke bottle etc., In order to be able to implement this feasibly on an economic scale, companies must and should adapt to the agile supply chain practices to quickly respond to the changing preferences of the customer.
- Ability to operate with least possible inventory: There is high risk of stock obsolescence and spoilage in maintaining high levels of inventory and not only that, it also freezes the capital base of the company which is much needed. One such industry where there's high risk in keeping high levels of inventory is, fashion industry.

In order to benefit and satisfy clients, an agile supply chain needs different identifying capabilities. It mainly involves responsiveness, flexibility and adaptability. The key components of an enterprise are agile, market-sensitive, process automation, network-based and virtual. They should be capable of being versatile, agile and adaptable to changing conditions in the market. This can be accomplished through collaborative partnerships, integration of systems, integration of information, and attention to customer/marketing to achieve customer-satisfied goals. These involve expense, time, competency and speed in the supply chain, contributing to the entire organization's competitive advantage.

4 How is Agile SCM Radically Different from Conventional SCM and Lean Concepts?

4.1 Conventional SCM

This Section will study about conventional SCM, agile SCM and lean concepts. The conventional SCM involves the following steps:

- Collection of raw materials: The first step involves the collection of raw materials which are required to make the final product.
- Collection of material from the suppliers: The manufacturers must collect all the required raw materials from the suppliers in order to produce the finished product.
- Manufacturing: The manufacturer then involves several processes and manufactures the final product from all the raw materials that have been collected.
- Distribution to the customers: This step involves distributing the finished goods to the customers or the end users.
- Consumption by the end customers: In this process, the customer or the end user purchases the finished goods and consumes it.

Nowadays technology has progressed than the traditional and conventional approaches are being challenged by modern ideas. With the same technical development, the SCM process has become fully digital, leading to a method called modern SCM. In developing countries, top freight shipping uses modern SCM, which is a testament to technical evolution and its impact on the logistics industry. It is very important to structure SCM not only around supply and demand, but also against forces and market dynamics on a larger scale for a business that is looking for growth and to optimize its profits. It is crucial to have a consistent and good understanding of the supply chain requirements of a company in today's environment where the economy is unpredictable, with limited profit margins and unexpected risks. More critical than ever, it is now.

4.2 Lean SCM

The lean supply chain is also named as the traditional "factory" chain. Their main focus is on producing high volume at low cost. The aim is to give consumers added value by reducing the cost of products and reducing waste. Instead of versatility and adaptability, this sort of supply chain management primarily focuses on the principles of durability and predictability. Instead of adjusting to the changing market conditions, development is planned well in advance, almost months or even years in advance. This pre-planning allows them to find the lowest possible cost for large quantities of items.

4.3 Agile SCM

Agile Supply Chain's main focus is on flexibility and adaptability to the changing situations. This is considered important for companies that wish to adapt to the unforeseen and unanticipated changes in economy, technology and demands of the customers. This SCM method also helps them in adjusting to the sourcing, sales and logistics of their organizations.

4.4 Agile Versus Lean SCM

Having to decide whether to implement a lean or agile SCM can often be as easy as using basic market analysis to assess the attractiveness of the product. In order to be competitive in the supply chain, companies should take a wider view of the market, taking into account the broader economic factors that are likely to impact consumer demand in the long run. Some common and important questions raised are:

- goods that are to be produced,
- target audience,
- consumer behavior,
- demand of the product being developed,
- market dynamics,
- fluctuations in economy,
- spending habits of consumers,
- partners of our supply chain and their operations.

For a specific company, we will determine which supply chain would be the best fit. It is possible to decide which supply chain strategy is right for you by considering the above questions. Fluidity and adaptability to react effectively to market changes are the distinction between lean and agile. The main focus of lean SCM is on reducing costs by manufacturing high quantities of low-variability goods. Probably, the focus of an agile SCM is mainly on reacting with smaller, flexible batches of goods to the consumer demand. A lean supply chain is also more predictable and cost-effective, whereas an agile supply chain is more versatile and adaptable.

5 Impacts of Digitization and Industry 4.0 on Agile SCM Manufacturing Process

The Digital Supply Chain (DSC) is an intelligent optimal system for digital hardware, software, and networks that supports synchronization to interact between organizations by making services more desirable, usable, and inexpensive with consistent, agile, and effective outcomes. It is based on massive data disposal, collaboration, and good communication [23]. Digitalization is the key here, as it provides both flexibility and efficiency. Apart from the seeming excitement around DSC, it is mostly debated from the perspective of customers.

In view of present unparalleled market instability and complicated customer needs, agile manufacturing is at the heart of gaining a sustained competitive advantage. In this aspect, agile manufacturing's emphasis on responsive adaptation would mitigate the destabilising effects of competitive demands on an organization's performance metrics [24]. Manufacturers have increasingly used technology, such as artificial intelligence, the internet of things (IoT), and predictive analytics, to drive success in their operations and supply chains in recent years.

The digitization and Industry 4.0 revolutions have the potential to affect all levels of company [25]. The deployment of SCM and its impact on corporate success and customer satisfaction are projected to have a significant impact on major industrial enterprises' success and consumer satisfaction. In recent years, digitalization in SCM has spawned a plethora of studies, one of which found that digitalization improves SC efficiency both internally and externally, as well as vertically and horizontally. Industrial digitalization can benefit from a variety of novel tools and approaches that have emerged as a result of technological advancements. Artificial intelligence (AI), the Internet of Things (IoT), cloud services, 3D printing, and sensors are just a few examples of such instruments [26–30].

SC Analytics, or Big Data analytics for SCM, is a component of the digital SC paradigm. These digital SCM domains are linked to and based on the described Industry 4.0 and Smart Manufacturing industrial digitalization principles [31] (Fig. 3).

Industry 4.0's objective is for enhanced digitization of processes. Traditional supply chains must evolve into a connected, intelligent, and highly efficient supply chain ecosystem. The existing supply chain consists of a succession of generally distinct, siloed steps that lead from marketing, product development, manufacturing, and distribution to the customer's hands.

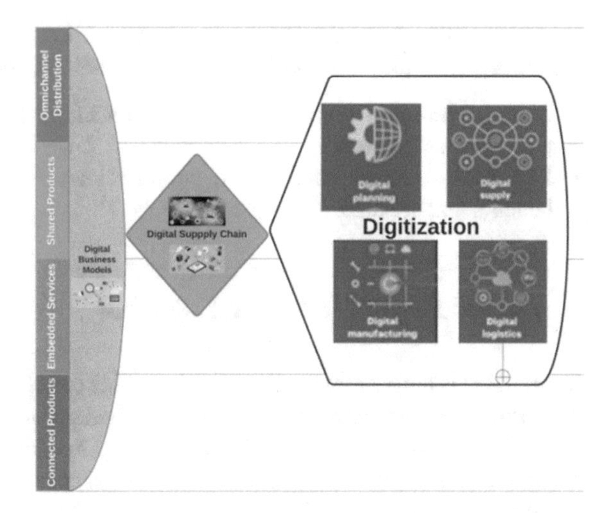

Fig. 3 Digitization supply chain

The influence of digitalization and Industry 4.0 on the supply chain (SC) ripple effect and disruption risk control analytics is investigated. The results from two separate domains, namely the impact of digitalization upon SC management (SCM) and the impact of SCM on ripple effect control, are combined in the research framework (Table 2).

Internet technology is more appealing to agile firms than Electronic Data Interchange (EDI). Despite the fact that EDI plays an important role in manufacturing, supplier and customer linkage, price quotes, and shipping notifications, manufacturers are more interested in the Internet, which may provide a more practical means for manufacturers and trading partners to communicate electronically [32–34]. Steel [35] anticipated that after 40 years or so of effort, there will be less than 100,000 EDI users worldwide. Because Internet commerce is still in its infancy, there is no specific definition for the function of the Internet in the supply chain in agile manufacturing. The process transforms digital inputs into value-added outputs [36]. The information is collected as raw material, and value-added information-based goods or services

Table 2 Transformation from Industry 1.0 to Industry 4.0 and digital ecosystems

Industrial revolution	Year	Descriptions
Industry 1.0	1800+	Mechanical production powered by water and steam
Industry 2.0	1900+	Machines production with electricity and combustion engine
Industry 3.0	1970+	Production by electronics, IT and robots
Industry 4.0	2015+	Production of digital products, services and business models by digital supply chain
Digital ecosystem	2030+	Virtualization process and industry collaboration as a key value driver

are created from the original raw data. The benefits derived from EDI use, according to Iacovou et al. [37], include both direct and indirect benefits.

- Reduced transaction costs or decreased inventory levels are direct benefits that are very easy to assess.
- Improved customer service and commercial partner connections are indirect benefits that are difficult to quantify.

6 Digital Impacts of Agile Logistics

Digitization is the act of gathering data from various silos or buckets and aggregating it to provide a comprehensive picture of individual operations from start to finish. The IoT idea has taken the supply chain sector by storm as the world continues to become increasingly linked; companies have reaped significant benefits from the implementation of a digital logistics [38]. The idea of digital logistics entails digitizing conventional data gathering, which is typically manual and prone to human mistake or delays, in order to improve and accelerate logistics processes, systems, and strategies.

All customers and stakeholders will have access to the information they need to make agile, informed decisions even in the midst of an unexpected situation if they use thoughtfully-integrated enterprise logistics applications that can collaborate and communicate effectively through a centralized information system. The digital transformation enables organizations to save money on operational expenditures, lost productivity, and precise order fulfilment [39, 40]. It improves data transparency for all suppliers, vendors, and partners engaged in supply chain management and operations.

6.1 Why Digital Logistics is Trending?

The internet is driving a demand for more speed and agility in the supply chain, which is why businesses are constantly searching for innovative methods to cut costs and improve communication between partners and service providers. As a result, supply chain management is shifting from traditional logistics to digital logistics. Digital logistics provide the agility that modern supply chain businesses require to effectively connect with each and every supply chain participant. Digital logistics is being adopted by modern supply chain organizations to improve distribution techniques such as technology, fulfilment data, and warehouse management.

6.2 *Impact of Digital Logistics on Business Operations*

Machine learning is used by the majority of digital logistics systems to make data-driven and important choices. This involves always striving for and driving supply chain process improvement. The supply chain team is no longer enslaved to the monotony of managing repetitive and time-consuming procedures by going through spreadsheet after spreadsheet in search of the information they require to take the appropriate action.

It may concentrate on strategically enhancing the process, which will have an immediate impact on customer satisfaction. As a result, this has an impact on the company's sales and profits. Digital logistics isn't a replacement for the human touch; rather, it's a tool that allows supply chain providers to focus on providing a better client experience.

In today's market, logistics is the newest brand differentiator. Even if all other aspects are identical, logistics may make the difference between returning customers, unfavorable reviews, and revenue lost. Consumers often anticipate a seamless experience from point-of-sale to delivery. Because of their greater reach, companies that embrace technology will always be ahead of the curve and prosper regardless of the challenges.

6.3 *Making the Switch to Digital Logistics*

Digital logistics is the quickest method to improve customer experience by providing clear, real-time visibility of order fulfilment, inventory status, and shipping updates. If you acquire access to more data, you should already have a strategy in place to make the most of it. Artificial intelligence and simplified procedures free staff from mundane duties, allowing them to focus on providing a better client experience. Consumer expectations will be driven by the digitization process, which will continue to evolve.

7 Conclusion

The logistics method for significant changes as a result of the advent of Internet technologies in business is the focus of this chapter. This isn't just about the digitization of some products that don't require transportation, but also about the fast-paced growth of e-commerce for other commodities. On the one hand, the Internet has removed supply chain intermediaries while also creating new sales and distribution channels. The ultimate consumer placing the order at any location and time has become the focal point of attention.

Home delivery has taken the place of the previous route to the store. Following the introduction of online sales, the client became an essential component of the logistics process and, in many cases, dealt with logistics services for the first time.

E-commerce is presently one of the most important drivers of economic growth. Because of reduced pricing, convenience, and a larger choice of items, an increasing number of people are ordering things through the Internet. The nature of supply chains in terms of its lean and agile approach in the context of the sort of product being moved (which can be sold via e-commerce), logistical solutions and supply chain management, and trends that form the image of supply networks are all discussed in this chapter.

References

1. Cohen, D., Lindvall, M., Costa, P.: An introduction to agile methods. Adv. Comput. **62**(03), 1–66 (2004)
2. Ashmore, S., Runyan, K.: Introduction to Agile Methods. Addison-Wesley Professional (2014)
3. Dingsøyr, T., Dybå, T., Moe, N.B.: Agile software development: an introduction and overview. In: Agile Software Development, pp. 1–13. Springer, Berlin, Heidelberg (2010)
4. Schmitt, A., Hörner, S.: Systematic literature review–improving business processes by implementing agile. Bus. Process Manage. J. (2021)
5. Shastri, Y., Hoda, R., Amor, R.: The role of the project manager in agile software development projects. J. Syst. Softw. **173**, 110871 (2021)
6. Seyedan, M., Mafakheri, F.: Predictive big data analytics for supply chain demand forecasting: methods, applications, and research opportunities. J. Big Data **7**(1), 1–22 (2020)
7. Handfield, R.B., Nichols Jr, E.L.: Introduction to Supply Chain Management. Prentice Hall, Englewood Cliffs, NJ (1999)
8. Kumar, K.: Technology for supporting supply chain management: introduction. Commun. ACM **44**(6), 58–61 (2001)
9. Ross, D.F., Weston, F.S., Stephen, W.: Introduction to Supply Chain Management Technologies. Crc Press (2010)
10. Janvier-James, A.M.: A new introduction to supply chains and supply chain management: definitions and theories perspective. Int. Bus. Res. **5**(1), 194–207 (2012)
11. Flynn, B., Cantor, D., Pagell, M., Dooley, K.J., Azadegan, A.: From the Editors: Introduction to Managing Supply Chains Beyond Covid-19-Preparing for the Next Global Mega-Disruption (2021)
12. Faugère, L., Lauras, M., Wamba, S.: (2021) Introduction to the minitrack on digital and hyper-connected supply chain systems. In: Proceedings of the 54th Hawaii International Conference on System Sciences, p. 2046
13. Fatorachian, H., Kazemi, H.: Impact of Industry 4.0 on supply chain performance. Prod. Plan. Control **32**(1), 63–81 (2021)
14. Myerson, P.: Lean Supply Chain and Logistics Management. McGraw-Hill Education (2012)
15. Arif-Uz-Zaman, K., Ahsan, A.N.: Lean supply chain performance measurement. Int. J. Prod. Perform. Manage. (2014)
16. Takeda-Berger, S.L., Tortorella, G.L., Rodriguez, C.M.T., Frazzon, E.M., Yokoyama, T.T., Oliveira, M.A.D.: Analysis of the relationship between barriers and practices in the lean supply chain management. Int. J. Lean Six Sigma (2021)
17. Zhao, P., Yin, S., Han, X., Li, Z.: Research on lean supply chain network model based on node removal. Phys. A: Stat. Mech. Appl. **567**, 125556 (2021)

18. Christopher, M.: The agile supply chain: competing in volatile markets. Ind. Mark. Manage. **29**(1), 37–44 (2000)
19. Oloruntoba, R., Gray, R.: Humanitarian aid: an agile supply chain? Supply Chain Manage. Int. J. (2006)
20. Croom, S., Baramichai, M., Zimmers, E.W., Marangos, C.A.: Agile supply chain transformation matrix: an integrated tool for creating an agile enterprise. Supply Chain Manage. Int. J. (2007)
21. Ahmed, W., Huma, S.: Impact of lean and agile strategies on supply chain risk management. Total Qual. Manag. Bus. Excell. **32**(1–2), 33–56 (2021)
22. Lavanpriya, C., Muthukumaran, V., Senthilkumar, K.M.: An effective model for vendor selection and allotment of order quantities in agile supply chain for multiple products and multiple suppliers in manufacturing industries. Solid State Technol. **64**(2), 743–761 (2021)
23. Büyüközkan, G., Göçer, F.: Digital supply chain: literature review and a proposed framework for future research. Comput. Ind. **97**, 157–177 (2018)
24. Gisbrecht, P.: Quantifying the impact of digitalization on manufacturing supply chain management (SCM) in a power generation company (Doctoral dissertation, Massachusetts Institute of Technology) (2018)
25. Garay-Rondero, C.L., Martinez-Flores, J.L., Smith, N.R., Morales, S.O.C., Aldrette-Malacara, A.: Digital supply chain model in Industry 4.0. J. Manuf. Technol. Manage. (2019)
26. DeVor, R., Graves, R., MILLS, J. J.: Agile manufacturing research: accomplishments and opportunities. IIE Trans. **29**(10), 813–823 (1997)
27. Houyou, A.M., Huth, H.P., Kloukinas, C., Trsek, H., Rotondi, D.: Agile manufacturing: general challenges and an IoT@ Work perspective. In: Proceedings of 2012 IEEE 17th International Conference on Emerging Technologies & Factory Automation (ETFA 2012), pp. 1–7. IEEE (2012)
28. Cao, Q., Dowlatshahi, S.: The impact of alignment between virtual enterprise and information technology on business performance in an agile manufacturing environment. J. Oper. Manag. **23**(5), 531–550 (2005)
29. Amjad, M.S., Rafique, M.Z., Khan, M.A.: Modern divulge in production optimization: an implementation framework of LARG manufacturing with Industry 4.0. Int. J. Lean Six Sigma (2021)
30. Onu, P., Mbohwa, C.: Industry 4.0 opportunities in manufacturing SMEs: sustainability outlook. Mater. Today: Proceed (2021)
31. Schrauf, S., Berttram, P.: How Digitization Makes the Supply Chain More Efficient, Agile, and Customer-Focused. Price Waterhouse Cooper (2018).
32. Iskandar, B.Y., Kurokawa, S., LeBlanc, L.J.: Adoption of electronic data interchange: the role of buyer-supplier relationships. IEEE Trans. Eng. Manage. **48**(4), 505–517 (2001)
33. Gunasekaran, A.: Agile manufacturing: a framework for research and development. Int. J. Prod. Econ. **62**(1–2), 87–105 (1999)
34. Centobelli, P., Cerchione, R., Ertz, M.: Agile supply chain management: where did it come from and where will it go in the era of digital transformation? Ind. Mark. Manage. **90**, 324–345 (2020)
35. Steel, K.: University of Melbourne. Australia, Private Commun. (1996)
36. Kalakota, R.: Debunking Myths about Internet Commerce. University of Rochester, USA (1997)
37. Iacovou, C.L., Benbasat, I., Dexter, A.S.: Electronic data interchange and small organizations: adoption and impact of technology. MIS Quart. 465–485 (1995)
38. Muh, F.N.: A framework supporting the design of a lean-agile supply chain towards improving logistics performance (2008)
39. Panigrahi, R., Borah, S., Bhoi, A.K., Ijaz, M.F., Pramanik, M., Jhaveri, R.H., Chowdhary, C.L.: Performance assessment of supervised classifiers for designing intrusion detection systems: a comprehensive review and recommendations for future research. Mathematics **9**(6), 690 (2021)
40. Chowdhary, C.L., Patel, P.V., Kathrotia, K.J., Attique, M., Perumal, K., Ijaz, M.F.: Analytical study of hybrid techniques for image encryption and decryption. Sensors **20**(18) 5162 (2020)

Supply Chain Management Using Blockchain, IoT and Edge Computing Technology

Arpit Jain and Dharm Singh Jat

Abstract Supply chain management is a backbone of the industry to serve the requirement of routine life. The supply chain process is complex as it involves several organisations across geographical boundaries. It involves various stakeholders with multivariant processes such as product information, manufacturers, logistics, consumers, product safety, etc. These are some centralised system's challenges. Blockchain technology can achieve transparency, decentralisation, and distributed trust to resolve these issues and improve the existing system. With the help of technology, managing goods, services, and information with an efficient system can be designed to achieve high performance. The health sector is a challenge for each country in this situation of pandemic Covid-19, and to perform the smooth operation, there is a need for an end-to-end tracking system, robust, secure, and strong framework. It is required to ensure the security and integrity of data with transparency in supply chain management. This study presented the conceptual framework of blockchain enabling supply chain management to provide the solution in the healthcare industry. This study provides an efficient way of utilising blockchain technology in the supply chain process and enhance the healthcare sector.

1 Introduction

In the supply chain industry, multiple end-users interact with the information, including the details of transactions. Due to the frequent data flow between end users, the system attracts attackers and to maintain the data integrity, and it is required to protect data. Various research has been done in the sector of securing information. Blockchain technology emerges as one of the promising techniques to ensure data security in the field of supply chain management. The edge computing drastically

A. Jain (✉) · D. S. Jat
Namibia University of Science and Technology, Windhoek, Namibia
e-mail: ajain@nust.na

D. S. Jat
e-mail: dsingh@nust.na

K. Perumal et al. (eds.), *Innovative Supply Chain Management via Digitalization and Artificial Intelligence*, Studies in Systems, Decision and Control 424,
https://doi.org/10.1007/978-981-19-0240-6_6

changed the scenario in terms of latency; through this, the computational latency and request-response with processing overhead is improved. The health sector has significantly improved with changing technology [1]. The cloud infrastructure is heavily utilised in the case of real-time applications and results in network congestion. The various challenge with the cloud to provide the efficient services and delay-sensitive response is not feasible due to presence of many stakeholders and the huge number of patients [2]. There is a requirement of the Internet of Things (IoT) enabled industrial elements of edge computing. The Edge IoT system will increase the bandwidth between cloud and edge devices by reducing latency, resulting in the quality of services [3].

1.1 Background

A. *Blockchain Technology*

Blockchain technology is decentralised distributed database management that can store all the transactions in the distributed ledger. The block generated is reflecting the transaction and append to the chain of blocks. These transactions are created by the parties or nodes involved in the system. The participants verify the transactions before adding to the chain [4]. The technology has an immutability feature that can ensure the security and integrity of data. With cryptographic hash functions, the information is stored in a tamper-proof manner to generate a trusted communication between end-users. This is achieved through a decentralised distributed ledger; the transactions are immutable once it is committed in the ledger [1]. Due to the inability of modification on data which results in transparency and security of data. This brings the smooth operation in supply chain management with ensuring the tracking and tracing of the transaction by end users. The health sector needs the support of technology to scale up the operation in the supply chain. With blockchain technology, the pharmaceutical industry will enhance the functioning of the supply chain from manufacturing to the consumer with information security.

As technology grows the trend of buying and consume products has been changed from traditional approach to the technological way [5]. The demand for customisable products and quick access are the necessity of the consumer; in that case, the efficient buying experience and transparency are the major requirements. Due to this requirement, new challenges and opportunities are automatically generated. The supply chain challenges are making the process smooth with data transparency, efficient network of participants, tracking and tracing of products, and optimised information availability.

B. *Forms of Blockchain*

There are three forms of blockchain which are public blockchain, private blockchain and permissioned blockchain. The public blockchains are open for all to participate and fully distributed such as Bitcoin, Ethereum etc. and uses the algorithm such as

Proof of Stake and Proof of Work. The Private blockchains enable the limited user who runs privately with verified participants in an organisation or private company. The private blockchain uses Distributed Ledger Technology (DLT). The permissioned blockchains are a combination of a public and private blockchain; in this, the customised network is designed to identify the participant based on the authentication process and assign the participants' roles. The example of permission blockchains are Ripple, R3's Corda and Hyperledger [6].

C. *Consensus Algorithms*

The consensus algorithms are designed to achieve the joint agreement on the data stored at the ledger by all the participants of the blockchain network to enable the distributed trust [7].

Through the consensus, the reliability in the network can be achieved within unreliable participants or nodes. Broadly, there are two categories of consensus algorithms: voting-based and proof-based algorithms [8]. The below Fig. 1 shows the classification of the consensus algorithm. With the mentioned consensus algorithm, reliability and trust are created in the blockchain network between participants. It depends on the requirement to choose the type of blockchain implementation. Bitcoin is a public blockchain, so the consensus algorithm like Proof of Work is implemented accordingly. Whenever there is a need at an organisational level to achieve the integrity and security of data, the private or permissioned blockchain type can be implemented. For supply chain system, the stakeholders are defined to participate, and there is no need for the public user to interfere. In this case, the permissioned blockchain would be more suitable. The end-users participating in the

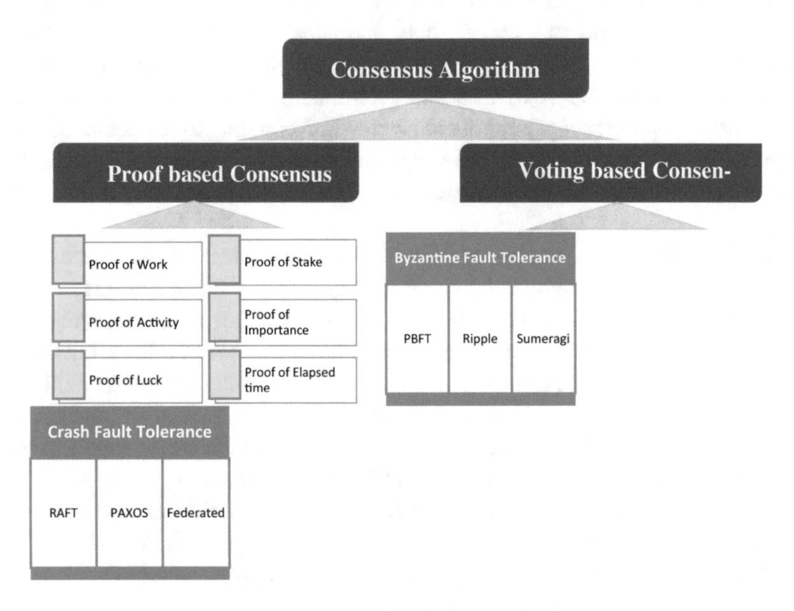

Fig. 1 Classification of consensus algorithm of blockchain technology

supply chain will be verified with the authentication process. Then the transactions are committed and agreed upon by all the participants in a close group. This will secure the operation of the supply chain as it is implemented with the permission or a limited number of participants, and it will reduce the complexity also. So in this way, the pharmaceutical industry can supply their drugs through multiple end-users named manufacturers, distributors, retailers and consumers without any implication security concerns.

2 Literature Review

Many researchers are studying to understand the use case of blockchain technology. Initially, blockchain technology was only known for the financial sector with named cryptocurrency. But with the several studies and technological improvements, it is found that blockchain technology is not limited to the financial aspects perhaps this technology can be scaled up in many industries like health care, insurance services, automotive, banking, and financial services, media and entertainment, government, telecommunications, travel and transportations and supply chain industry to achieve the security, transparency, and traceability where multiple stakeholders involved. A detailed literature review has been done to better understand the depth of blockchain technology in various sectors and the supply chain industry.

The study shows a comprehensive review of blockchain-based supply chain management with its capability, benefits, challenges, potential usage, and future scope. The study highlighted the major concerns: end-user participation and association, transparency and traceability, data digitalisation with integration of supply chain, and a normal framework for a blockchain platform. This study analyses various techniques to improve trust and performance in a distributed environment with future impacts on supply chain management. According to research, most of the studies followed the blockchain-based system design, the experimental aspect with simulation to evaluate the performance and effectiveness of the supply chain sector [9].

The study shows the importance of blockchain technology in the agriculture and food supply chain. It provides the solution using blockchain technology based on the Ethereum network to deploy the smart network. With this, data integrity can be ensured due to the immutable behaviour of blockchain technology, but there are some challenges with blockchain, like traceability, accountability, and credibility of stakeholders. The study proposes that the transactions are stored on Interplanetary File Storage System (IPFS) in form of blocks that return the stored data's hash that ensures the security and integrity of data. The experimentation was done with a smart contract and a designed algorithm to analyse the vulnerability and security of the domain. The study contributed to the reputation system to maintain the credibility of entities and the quality scale of the product in the supply chain [10].

The study proposes new features and understanding to improve blockchain implementation in the industry. It shows the different perspectives from the academic background and business point of view in blockchain technology. The study provides five enablers that are (i) Access enabler which includes identity and digital signatures, (ii) Value creation enabler, which deals with Artificial Intelligence and data; (iii) Interoperability enabler, which provides tokenisation, (iv) Remote enabler with the Internet of Things; and (v) Social enabler, which deals with blockchain technology in Supply chain management. These enablers will conceptualise the framework required in the industry for blockchain implementation. With the analysis of blockchain technology, the study found that distributed trust will sustain innovation in supply chain management [11].

In agriculture, the supply chain management process plays a significant role; if it is not maintained well with smooth operation, there would be a decay of farming items. It is observed that the proper supply chain management process with verification at various stages are required. To provide transparency in the system between consumer and producer a blockchain technology is required to implement. The study proposed a blockchain-based solution to record the process of the supply chain in the agriculture sector. All the data related to farming goods like quantity, type, price etc., are stored on the blockchain, and the collected data will be specified using the smart contract. The committed transaction is now added to the block. The records about the agricultural goods are now updated on the distributed ledger. Through this, the IoT-based blockchain system process will be transparent and immutable as no one can modify data, ensuring the security and integrity of data. With this technology, the life of farmers can be improved, where the status of goods can be easily viewed through the supply chain management process equipped with blockchain technology [4].

The study shows the consequences of blockchain technology in the retail industry of business. It analyses the process of the companies to improve the customer base in the retail sector with blockchain technology. Due to decentralised distributed trust in the supply chain process, the outcomes are outstanding in cost reduction and transparency in financial transactions. The study highlighted that loyalty and reward programmes with transparency can preserve the customers, but there are certain challenges at an implementation level. The evolution of blockchain in retail industry with secure ledger, smart contract to avoid the intermediatory which causes the smooth and cost effective operation [12].

The study proposes the Ethereum based blockchain approach with distributed decentralised ledger for product traceability. The supply chain in the healthcare sector is a kind of mesh network with multiple organisations and stakeholders to provide services. Due to complexity in the structure, lack of security, inaccurate information and transparency issues are always challenges. There is a need for an end-to-end tracking system that can ensure product safety, transparency, and traceability with transaction data storage. The study proposed the system architecture and algorithm to ensure the characteristics of blockchain technology with immutable transaction history with a security analysis to enhance the supply chain in the pharmaceutical industry. Many stakeholders like manufacturers, distributors, pharmacies, and patients participate in the smart contract in this architecture. The study proposed two important features; one is the prevention of manipulation of data as blockchain provides immutability, secondly it eliminates the failure at any single point. The study shows some limitations in the health sector supply chain that is efficiency, interoperability and scalability. With the evaluation of the developed blockchain model in drug supply, it is found that the solution is cost-effective and ensures security and data integrity [13].

The study shows the analysis of blockchain implementation with the Internet of Things in the pharmaceutical industry. The research shows the importance of IoT to track the supply of products from one stakeholder to another. The sensors help collect the data from the process and track the RFID chips that can be attached to the product supplied. The combination of blockchain and IoT in supply chain management will improve the system to ensure security and data integrity, distribution and logistics, assets and tracking management with data immutability. The study found some challenges regarding standardisation, scalability, interoperability and sharing of data. The study concluded with future potential in the supply chain with blockchain technology to gain a complete view of the pharmaceutical industry's product manufacturing, supply, and consumption [14].

The study proposed a blockchain-based Medicine Supply Chain Management (MSCM) system implemented in the distribution of medicines in Indonesia within the public health sector. To analyse the performance JMeter tool is used to measure the throughput of transactions, latency and the availed resources. A case study was presented in the medical sector for tracking and tracing the supply of medicine between various stakeholders. The experimentation was done on permissioned blockchain with the help of two HTTP methods, GET and POST, using two independent variables, the type of request and number of clients. The study concluded that the performance decreases due to an increase in the number of clients, which means latency response increases. The proposed model provides better efficiency in the medical or drug supply chain [15].

The study shows the opportunities to implement blockchain technology in various industrial sectors to resolve issues like product supply information, transactional data, lifecycle of product from manufacturing to consumer, transport history, etc. The study analyses the characteristics of blockchain technology such as decentralisation, the immutability of data and distributed trust to propose the conceptual framework in supply chain management. The preliminary study focused on the suitability of blockchain in supply chain management stages such as fulfilment of orders, supplier information and relationship, manufacturing process, and product demand. In this work, the evaluation is done on the suitability of blockchain technology on supply chain management in terms of the decentralised trust to mitigate the issues of distrust, transparency and sharing the transactional information. Further, the remaining stages of supply chain can be explored to evaluate the performance in the integration of blockchain algorithms [16].

The study proposed the framework to show the improvement in the quality of the supply chain management process. The framework includes the IoT layer where GPS, Sensors, RFID and barcodes are utilised to collect the information related to location, assets, logistics and transactions. The data layer includes the distributed ledger which stores the collected data into the ledger. The third layer is contract layer which includes digital identity, data sharing and quality control. The fourth layer is a business layer which enables the enterprises to take decisions, monitoring the manufacturing activities and control the supply chain processes. The further study evaluated the real-time quality monitoring and control, contract automation process, logistics planning and loan investigation assistance in supply chain management. The study concluded the efficient framework for quality management in the supply chain using blockchain technology [17].

The study proposes the CLOUDITY using blockchain technology and selective market, a cloud-based supply chain management system. This system has solved various cases, such as ensuring security with authentication, service selection, resource provisioning, and access control between multiple stakeholders. The proposed solution modified the cloud-based services to create the latest broker architecture known as Selective Market (SELAT). The user request is directed through a particular market that recommends Cloud Service Platform (CSP). Out of several CSPs, the user selects one to avail the resources. With this system, the information can be shared between companies and manufacturers more securely with the help of blockchain technology as blockchain technology provides immutability, which can improve the integrity and security of data against modification, further improving the performance with less processing time [18].

The study analyses the challenges such as inaccurate information in maritime transportation. The study proposes a digital identity management mechanism to avoid data inconsistency in a blockchain-based cargo management system based on challenges. To secure the information, the participants involved in cargo management must be vetted and set up their public–private key pair infrastructure to verify the identity while accessing and sharing the information. The root node can verify the public key of participants and government agencies with the certificate authority to identify the identity digitally. The study concluded that blockchain plays an important role in preventing the cargo inspection process and loss of information; the designed model bridges participants' physical and cyber presence to reflect the accurate information, proper status of the cargo [19].

The study shows the implementation of blockchain technology to enhance the verification process of food quality and the flow of food products in agriculture supply chain management. The public blockchain has been implemented to enable transparency within participating users in the network. The peers of the blockchain system validate the transaction involved with the instance creation of a smart contract, and the information related to a food product is deployed on the blockchain network. The study highlighted the token-based scheme which reflects the value of the product, and as the flow goes forward, the ownership gets updated according to the transaction. Peers will assign the reputation form of token based on the certification request for the specific product to the farmers. The product identification is made based on a unique code is known as a quick response (QR) code. The study shows an efficient way to improve the transparency between the farmer as producer and consumer. The evaluation and testing of the transaction were done with the help of the Ganache network and the information stored on the blockchain network, which results in better transparency and trust between various stakeholder in the field of agriculture supply chain management [20].

The study proposed the Intelligent Decision Support System (IDSS) design to create computer expertise to make decisions based on knowledge base components. The study highlighted few challenges in the sugarcane supply chain industry: mistrust between stakeholders, incomplete transactions of data, and traceability of the supply process with financial information. To resolve the mentioned issues, the study uses Artificial Intelligence techniques with blockchain technology. With the help of IoT, the supply chain flow can be easily tracked and enforce the visibility, traceability, and transparency of transactions using blockchain mechanisms. The study concludes the efficient design which can improve the sugar agroindustry to manage the resources and financial aspect with ensuring the security and integrity of data [21].

The study shows the developed application to enhance supply chain tracking based on Etherium blockchain technology. It enables various participants to utilise the independent hardware platform approach with the decentralised application (Dapp). The study shows the experimentation of designed applications with two fully operational blockchain cases; the first is interaction with the producer, which will add product details, select the products, and add the producer's license information, certificates, and profile. Secondly, the set of products are connected through a chain generated with the help of QR codes which results in the verification to ensure data integrity [22]. The study shows the different perceptions in adopting the blockchain from several motivators and barriers in organisations from various sectors. It is highlighted that the blockchain was primarily popular for the financial aspect in terms of cryptocurrencies, but practitioners found that it is not limited to the financial industry it can be scaled to multiple types of agreements, contracts, tracking and exchange in different sectors.

With the analysis of various end-users, it is found that there is a lack of expertise in technology. If we adopt it, it will increase and enhance information security. According to the study, the industry is embracing the blockchain technology, many companies and organisation have potential to implement their supply chain process with decentralised distributed trust [23].

3 Conceptual Framework

The conceptual framework may provide the supply chain management using blockchain technology is a concrete solution for achieving transparency, traceability and immutability within various stakeholders. To save time and improve the efficiency of the supply chain process, edge computing brings cloud resources to the edge of the network. The Internet of Things involved collecting data such as scanning QR codes, barcodes which can further help track the product supply from one stakeholder to another. The edge layer reduces the burden of the cloud as all the request response will be handled at the edge. The end-user participating in the supply chain such as manufacturer, distributer, supplier, retailer, consumer, etc. is accessing the distributed ledger at cloud and committing the transaction from their end. The information stored by the end user are mainly product details, supplier details, consumer details, retailer details, prices, taxes information, and validity. To secure this information, blockchain technology is implemented, which can be stored on a decentralised distributed ledger. All the transactions are committed to building trust through smart contracts.

Figure 2 shows that the real-time data will be collected from sensors, IoT devices and filtered at the Edge layer. Further, this information will be stored in the cloud using blockchain technology as historical data so that no one can modify that data. The data will be held in the form of blocks using a cryptographic hash function. The supply chain participants will access this data through an authentication mechanism and append the block of data for additional information; this is how the chain of blocks will increase in the form of immutable data shared through all the participants.

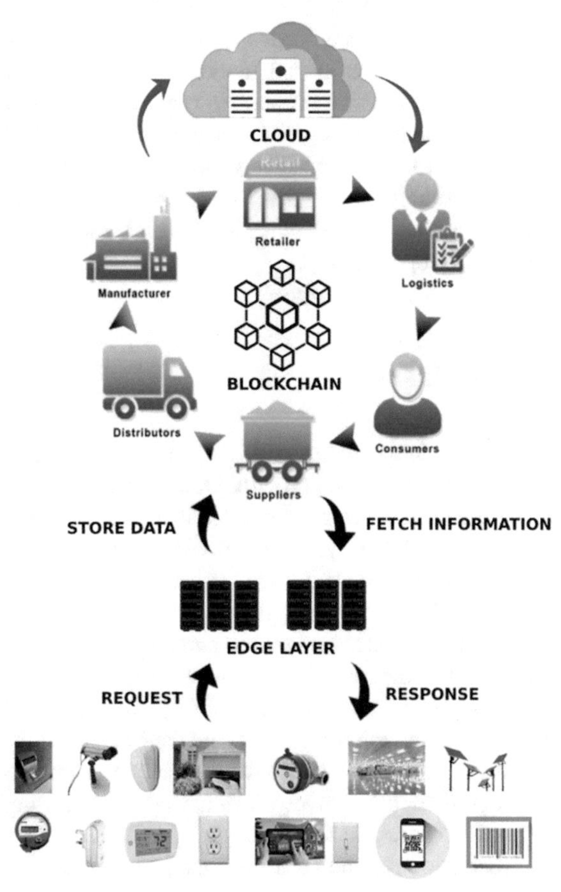

Fig. 2 Conceptual framework of blockchain-based supply chain management

The supply chain process from manufacturer to consumer will generate a huge amount of data. The major challenge in the supply chain is multiple stakeholders generate data from their own end. This makes the process complex due to distributed ledger it requires heavy computational machines. Somehow it is better than the traditional approach of centralised database, which has many disadvantages such as concurrent access, network congestion, security issues and misuse of paper-based certificates. The blockchain and IoT bring a rapid change in the process from production to product delivery to customers.

With the technological revolution, the supply chain industry is completely changed from the traditional process. The best example is an eCommerce website, where users can place an order from anywhere without stepping out, and the product is delivered at the doorstep. In the eCommerce industry, the supply chain plays a vital role in product information, comparative analysis, placing the order, distributing from one end to another with the complete tracking of delivery to the consumer. The process is smooth, but there are certain challenges with the security of the information, identity, and correctness of the ordered product. All the information is stored at the cloud, which makes a burden on the cloud to access the information and maintain the security. The proposed model resolves the challenges with the help of edge computing to provide immediate requests and responses between user and cloud. To ensure the integrity and security of the data, blockchain has been implemented. Various end users participating in the supply chain can contribute securely, and verification with decentralised trust can improve the system. The computational cost, bandwidth and storage will also be optimised, and the processing power can be increased proportionally.

4 Conclusion

The IoT devices, sensors and RFID technology play an important role in collecting the data and the digital growth. Due to this, supply chain management is error-free and digitalised to store valuable information from many stakeholders. Further, with the help of technology, it is required to ensure the security and integrity of the information. Blockchain technology solves this issue of security and information sharing by storing the data in a distributed ledger that can be shared with the involved participants. The data is stored on the block with a cryptographic hash function to ensure security. The data block can not be modified by any of the participants involved, bringing transparency and traceability. The proposed conceptual framework based on blockchain technology implemented on the cloud to access by the end user. It is difficult for the delay-sensitive request to access the cloud resources, and hence the edge computing layer plays a vital role in bringing the cloud resource to the edge of the network. Due to edge layer, the data from IoT devices and sensors will be processed fast, and an immediate response can be achieved. Further, this data is stored on the blockchain, which is verified by all the participants in the supply chain management. With this model, the supply chain process will be enhanced to be efficient and transparent with the immutability of data. In our analysis, we found some challenges as the chain of blocks grows, the system's complexity will be increased. To overcome the challenges of complexity in the supply chain, the hybrid blockchain or permissioned blockchain can be implemented. So that confidential data related to the products, suppliers, manufacturing, and digital identity with tracking can be placed or stored on the blockchain, and other data can be off the chain. The data stored on the blockchain will be verified and approved by the participant using a smart contract and consensus algorithm. This will bring confidentially, integrity and availability to the supply chain management process.

References

1. Jain, A., Jat, D.S.: An edge computing paradigm for time-sensitive applications. In: Proceedings of the World Conference on Smart Trends in Systems, Security and Sustainability, WS4 2020, pp. 798–803 (2020). https://doi.org/10.1109/WorldS450073.2020.9210325
2. Pace, P., Aloi, G., Gravina, R., Caliciuri, G., Fortino, G., Liotta, A.: An edge-based architecture to support efficient applications for healthcare industry 4.0. IEEE Trans. Ind. Inf. **15**(1), 481–489 (2019). https://doi.org/10.1109/TII.2018.2843169
3. Pratim, P., Dash, D., De, D.: Journal of network and computer applications edge computing for internet of things: a survey, e-healthcare case study and future direction. J. Netw. Comput. Appl. **140**(March), 1–22 (2019). https://doi.org/10.1016/j.jnca.2019.05.005
4. Sudha, V., Kalaiselvi, R.: Blockchain based solution to improve the Supply Chain Management in Indian agriculture. In: International Conference on Artificial Intelligence and Smart Systems (ICAIS-2021). pp 1289–1292 (2021). https://doi.org/10.1109/ICAIS50930.2021.9395867
5. Chowdhary, C.L.: Growth of financial transaction toward bitcoin and blockchain technology. In: Bitcoin and Blockchain, pp. 79–97. CRC Press (2020)
6. Sangeetha, A.S.: Blockchain for IoT enabled supply chain management—a systematic review, pp. 48–52 (2020)
7. Gupta, C., Mahajan, A.: Evaluation of proof-of-work consensus algorithm for blockchain networks. In: 2020 11th International Conference on Computing, Communication and Networking Technologies, ICCCNT 2020 (2020). https://doi.org/10.1109/ICCCNT49239.2020.9225676
8. Alsunaidi, S.J., Alhaidari, F.A.: A survey of consensus algorithms for blockchain technology. In: 2019 International Conference on Computer and Information Sciences, ICCIS 2019 (2019). https://doi.org/10.1109/ICCISci.2019.8716424
9. Chang, S.E., Chen, Y.: When blockchain meets supply chain: a systematic literature review on current development and potential applications. IEEE Access **8**, 62478–62494 (2020). https://doi.org/10.1109/ACCESS.2020.2983601
10. Shahid, A., Almogren, A., Member, S., Javaid, N., Member, S., Al-zahrani, F.A., Zuair, M., Alam, M.: Blockchain-based agri-food supply chain: a complete solution. IEEE Access **8**, 69230–69243 (2020). https://doi.org/10.1109/ACCESS.2020.2986257
11. Valle, F.D.: Blockchain Enablers for supply chains: how to boost implementation in industry **8** (2020). https://doi.org/10.1109/ACCESS.2020.3038463
12. Hader, M.: Blockchain technology in supply chain management and loyalty programs: toward blockchain implementation in retail market, pp. 2–4 (2020)
13. Musamih, A., Salah, K., Jayaraman, R., Arshad, J., Debe, M., Al-Hammadi, Y., Ellahham, S.: A blockchain-based approach for drug traceability in healthcare supply chain. IEEE Access **9**, 9728–9743 (2021). https://doi.org/10.1109/ACCESS.2021.3049920

14. Premkumar, A., Srimathi, C.: Application of blockchain and iot towards pharmaceutical industry. In: 2020 6th International Conference on Advanced Computing and Communication Systems, ICACCS 2020, pp. 729–733 (2020). https://doi.org/10.1109/ICACCS48705.2020.9074264
15. Kumiawan, H., Kim, J.Y., Ju, H.: Utilisation of the blockchain network in the public community health center medicine supply chain. In: APNOMS 2020–2020 21st Asia-Pacific Network Operations and Management Symposium: Towards Service and Networking Intelligence for Humanity, pp. 235–238 (2020). https://doi.org/10.23919/APNOMS50412.2020.9237042
16. Yousuf, S., Svetinovic, D.: Blockchain technology in supply chain management : preliminary study, pp. 537–538
17. Chen, S.: The Fourteenth IEEE International Conference on e-Business Engineering A Blockchain-based Supply Chain Quality Management Framework (n.d.). https://doi.org/10.1109/ICEBE.2017.34
18. Asyrofi, R., Zulfa, N.: CLOUDITY: Cloud supply chain framework design based on JUGO and blockchain. In: 6th Information Technology International Seminar 19–23 (2020). https://doi.org/10.1109/ITIS50118.2020.9321013
19. Xu, L., Chen, L., Gao, Z., Chang, Y., Iakovou, E.: Binding the physical and cyber worlds : a blockchain approach for cargo supply chain security enhancement, pp. 1–5 (2018)
20. Basnayake, B.M.A.L., Rajapakse, C.: A Blockchain-based decentralised system to ensure the transparency of organic food supply chain, pp. 103–107 (2019)
21. Ekawati, R.: Design of intelligent decision support system for sugar cane supply chains based on blockchain technology, pp. 153–157 (2020)
22. Niya, S.R., Dordevic, D., Nabi, A.G., Mann, T., Stiller, B.: Blockchain-based supply chain tracking, pp. 11–12 (2019)
23. Polytechnic, W., Polytechnic, W., Polytechnic, W.: Blockchains and the supply chain: findings from a broad study of practitioners. **47**(3), 95–103 (2019). https://doi.org/10.1109/EMR.2019.2928264

COVID-19 Outbreaks Challenges to Global Supply Chain Management and Demand Forecasting on SCM Using Autoregressive Models

Mansi Chouhan and Devesh Kumar Srivastava

Abstract Global-SCM (Global-supply-chain-management) is characterised as the distribution of goods and services to maximise profit and reduce waste within the global network of a transnational business. The goal of this chapter is to provide information on the challenges faced by global supply chain management as a result of the Covid-19 pandemic, and to suggest solutions to these challenges through machine learning and artificial intelligence. The work presented in this research constitutes a contribution to modelling and forecasting the demand in a retail sales company, by using time series approach. Our work shows how the historical data of retail items could be utilized to forecast future demands and how these forecasts affect the supply chain management. The historical demand information was used for forecasting future demands using several autoregressive integrated moving average models like AR, MA, ARMA, ARIMA, and SARIMA. By comparing their Akaike's Information Criterion (AIC) values we get to know that ARIMA model is best suited for demand forecasting of the current retail item sale data. The result from the proposed work so obtained proves that the proposed model could be utilized for demand forecasting to meet the future demands in the retail sales item company.

Keywords COVID-19 · Coronavirus · Industry-4.0 · Epidemic outbreak · Supply chain · Retail industry · Globalization · Logistics · Manufacturing · Procurement · Sourcing · Supply chain management · Demand forecasting · Time series · Autoregressive integrated moving average (Arima)

1 Introduction

Global supply chain management which is abbreviated as Global-SCM handle's the supply chain management at global level where it is linked to several networked businesses and distributes goods and services to different companies around the globe to maximise the profit and reduce the waste [1].

M. Chouhan · D. K. Srivastava (✉)
Department of Information Technology, Manipal University Jaipur, Rajasthan 303007, India

K. Perumal et al. (eds.), *Innovative Supply Chain Management via Digitalization and Artificial Intelligence*, Studies in Systems, Decision and Control 424,
https://doi.org/10.1007/978-981-19-0240-6_7

Handling Global SCM becomes very difficult in current times with the outbreak of Covid-19. The pandemic generates many challenges and complex issues in various operations of Global SCM like import, export, manufacturing, procurement and distribution of goods and services. For handling global SCM in the current scenario we can use the time series models to conduct demand forecasting on historic data so that the retail company owner knows beforehand that how much supply they needed to meet the demands of the products in upcoming future [2].

1.1 Coronavirus Pandemic and Its Impact on Supply Chain Management

Corona viruses or COVID-19 are a large group of viruses which are the reason to cause mild respiratory tract infections in humans, varying from common cold or can be more severe like SARS (Severe_Acute_Respiratory_Syndrome) and MERS (Middle_Eastern_Respiratory_Syndrome. Corona virus comes from the family of RNA viruses in the Nidovirales order.

Despite numerous supply-chain disruptions caused by disasters over the last decade, such as a volcano eruption in Iceland, the Japanese earthquake and tsunami. Most businesses were caught off guard by the Covid-19 pandemic. The Covid-19 outbreak in China, stated they were still in data collection and evaluation mode, manually trying to figure out which of their suppliers had a site in the restricted areas of China. There are several causes for this issue, as well as potential solutions.

1.2 What is Global SCM?

The term "global supply chain management" is synonymous with "supply chain management," however it focuses on transnational firms and organisations. The advantages of Global SCM are Higher_Efficiency_Rate, Inventory_Buffers, Optimal_Shipping_Options, Mitigate_Your_Risks, Stay_On_Top_Of_Demand, Eliminate Waste, Minimize Delays, Improve Customer Service, Reduce Your Overhead Costs [3].

1.3 Why Demand Forecasting Using Auto Regressive Models?

Demand forecasting is very important in inventory module of supply chain management. Inventory stock levels depend on demand's forecasts. In reality, we can state that erroneous demand estimates can result in large expenses and material waste, showing that the process is not improving. Effective production plan can be created based on precise projections in order to reduce total production costs from procurement, processing, storage, and distribution. Smaller supply chain costs, lower inventory, higher customer satisfaction, higher return on assets, and shorter lead times were all expected profits from these estimates. However, this optimal creation of plan should meet different company constraints such as minimum production lots, production capacity, and so on [4].

A popular and widely used ARIMA model is the statistical method for time series forecasting. Autoregressive models are the most widely used approaches for time series forecasting and provide complementary approaches to the problem. Due to covid-19 pandemic there were many problems faced by the retailers. To ease their problem we conduct demand forecasting using time series data of a retail store where the historic sale data of the retail items sale is used to generate the future demands for the retail items so that the retailer know how much supply they need to meet the demands in the future. This demand forecasting technique, which we will demonstrate in the proposed study, will employ the most well-known time series models, such as AR, MA, ARMA, ARIMA, and SARIMA [5].

Autoregressive models operates under the proposition that past values have an effect on current values, which makes the statistical technique popular for economics, analyzing nature, and other processes that vary over time. Multiple regression models forecast a variable value of predictors using a linear combination, whereas autoregressive models use a combination of past values of the variable. Time series is used because it generates the predictions for the nearer future demands using past data. In this covid-19 scenario we need to generate the predictions more frequently for example after every week to get more accurate results for the upcoming future needs. Time series will be best suited for predicting nearer future forecasting.

We use autoregressive models and a time series technique to forecast demand in this proposed work. We used time series because we were creating forecasts based on data collected across time, and autoregressive models because we were making predictions based on data collected in the past. So by using past data we will predict the upcoming future demands for retail items so that it would help the retailers to meet the future demands and also without any wastage of resources and materials [6].

2 Literature Review

The digitization of the production process is a necessity for today's industry. The manufacturing industries are currently evolving from mass production to customised production. Growing productivity is driven by recent developments in manufacturing technologies and industrial applications. The term Industry 4.0 stands for the 4th industrial revolution, described as a new level of organization and control over the life cycle of the entire value chain of products. It includes the Internet of Things, Smart Manufacturing, Industrial Internet, and Cloud-based Manufacturing. Industry

4.0 requires the strict incorporation of humans into the production process in order to continuously improve and focus on activities that add value and reduce waste [7].

The author emphasis on supply chain disruption upstream. The system develops and evaluates a model that forecasts the frequency of supply chain disruptions based on a multi-dimensional conceptualization of upstream supply network complexity. The three dynamic drivers analysed, namely horizontal, vertical, and spatial complexity drivers, all enhance the frequency of disruptions, but they also combine and intensify one other's impacts in a synergistic manner, according to empirical findings [8].

The author focus on supply chain management strategy and supply chain performance practises in supply networks. The primary data gathering technique was a questionnaire that was organised by job title and function. The response rate was 62 percent, with 51 percent of the questionnaires being usable. The sampling was done for the sake of convenience. The results were assessed using the mean standard deviation and correlation between the independent and dependent variables. The findings show that supply chain management practises have a statistically significant relationship with supply chain performance [9].

Blockchain is a digitalization driver in supply chain management, with researchers utilising transaction cost theory to obtain early insights into how blockchain can affect future supply chain interactions. There were two major findings reached. To begin with, blockchain cuts the transaction and governance costs of supply chain transactions dramatically. Second, even under limited settings, the blockchain-based economy pushes many transactions into more market-oriented governance structure [10].

Industry 4.0 also termed as the 4th industrial revolution, is a collection of advanced manufacturing and information technologies designed to meet the individualized needs of various human areas in less time. All technologies in Industry 4.0 are interconnected. The medical stakeholders collaborate each other for vaccine production and use, healthcare equipment and logistics, check-ups, monitoring, identification, and determining appropriate behavior with minimal human physical intervention. The data captured by advanced technology provides accurate updates on the gathering of people [11].

Machines in Industry 4.0 factories are assisted by wireless networking and sensors. These sensors are connected to a network that can see and track the entire production line as well as make decisions on its own. To address the COVID-19 pandemic shortage, Industry 4.0 employs smart manufacturing processes to produce critical disposable products. During this crisis, it provides a smart supply chain of medical disposables and supplies, ensuring that patients receive the requisite medical products on time [12].

Supply Chain Industry 4.0 will meet the needs for customized face masks and gloves, as well as collect data for healthcare systems to properly monitor and handle COVID-19 patients. With proper surveillance systems, it is useful to provide day-to-day updates of an infected patient, area-wise, age-wise, and state-wise. We also agree that proper application of these innovations would assist in enhancing public health education and communication. These Industry 4.0 innovations have the ability to produce a slew of new concepts and strategies for combating local and global medical issues [13].

Studies show that the COVID-19 corona virus disease-free equilibrium (E0) does not follow the local or global asymptotic stability criteria. This means that the corona virus COVID 19 has still not been cured by the pandemic as declared by WHO (2020) and precautionary measures are recommended by quarantine and observatory measures [14].

With the world-wide outbreak of COVID-19, a precise model is proposed to make prediction on how the corona-virus pandemic will evolve. It becomes important and urgent to assist the policy makers in various countries to address the Epidemic Outbreak and Determine Policies to manage Spread more efficiently and effectively [15].

Investigating whether the weather has any role in Spreading the COVID-19 and the way that Knowledge are often accustomed arrest this fast Spreading virus/disease. It highlights that the humidity and temp both are extremely significant for transmitting the virus but the stronger factor is temperature. Due to this the mapping of supply networks can be time-consuming and challenging [16].

Multiple regression models have been defined with the goal of forecasting a variable of interest using a linear combination of predictors. In particular, in an auto-regression model, the variable of interest is forecasted using a linear combination of its past values (the term auto-regression indicates that it is a regression of the variable against itself), while a moving average model uses past forecast errors in a regression-like model. Sometimes, as a preliminary step to the regressive analysis, time series need a differencing transformation to stabilize the mean of a time series and so eliminating trend and seasonality. A combination of differencing, auto-regression and moving average methods is known as AutoRegressive Integrated Moving Average model (ARIMA) [17].

3 The Challenges in Different Aspects of Global SCM Were as Follows

3.1 *Distribution of Goods and Services*

(a) Challenges that are unique in nature may arise from the distribution of different products, for e.g. Staffing of warehouses is a need for direct distribution, responsiveness and smarter allocation across the channels.

(b) Retailing is also having an effect on the pandemic in a peculiar way. In the current times of lockdown and curfew scenarios around the world, there is a tremendous demand for essential goods rather than the luxury products by which the retailers suffered from a huge loss and some of them have already put their shutter down while many have been severely challenging on operating margins and models.

(c) On the other hand, if we're talking about the consumer side, where consumers get panicked and do stuff like hoarding/stocking essential commodities and taking over-the-counter medications. To cope up with the pandemic that has led to unusual stress on supply chain management structures. It is not unusual for consumers to panic and create a stockpile of food and other essential commodities during times of crisis. But if it goes beyond a few weeks of stockpiling, this will lead to a situation of stress. This causes an unnatural rise in demand and supply, and produces large fluctuations that are extremely difficult to handle, which together produce a bullwhip effect that often leads to artificial shortages in the supply chain [18].

3.2 Manufacturing Goods and Services for SCM

(a) **Import**: In order to produce a single final product, it is important to import various raw materials or semi-finished/finished components from different companies within the country or from different countries to manufacture goods and services, and sometimes even from different continents, to be shipped to the producer or to the consumer.

However, due to lock-down or curfew situations in different regions or countries, the availability of raw materials or semi-finished/finished components for importing products and services is challenging and often more complicated at the present time [19].

(b) **Export**: Similarly, there is a significant dependency on logistics for exporting various products that are used for importing and exporting different goods and services, which in turn causes a disruption to the management of the supply chain. There is also a challenge in exporting products and services to the regions that had lockout or curfew like circumstances.

(c) **Cost factor**: Imports and exports in manufacturing industries are becoming costly as a result of a pandemic, for example, goods with low cost semi-finished or finished products originate from countries such as China, where there is a risk of getting sickness from products and goods, which in turn causes the manufacturer to buy raw materials from various suppliers at high cost and the cost of shipping products is also high.

(d) **Man Power**: The machines in the industries need human power to initiate production to produce the final products, which is hard to get in the current scenarios. Because there's more to the fear of losing lives than losing a job.

3.3 Procurement of Goods and Services for SCM

The purchase of goods and services from pandemic-affected nations, the risk distribution becomes more apparent. As a result, purchases of goods and services from these countries are curtailed, affecting the country's economic situation.

4 Suggesting Solutions to the Challenges, (Crucial for Future SCM)

(a) **Supply chain control tower**: A supply chain control tower is not a physical tower or object; it is a centralised hub containing all the necessary technology, organisational instruments, and processes to capture data from all stages of the supply chain, from the manufacturer to the consumer.

(b) **Predictive analytics** is to guide the network towards greater performance and improved processes, **whereas** the control tower uses real-time data to identify weaknesses and strengths within the supply chain.. Throughout a complex supply chain network, supply chain control towers allow you to operate in more smarter and efficient way. In doing so, you provide your customers with greater benefits while reducing costly exceptions [20].

From procurement to delivery, it is a single source of truth for all trading partners to see and adapt to changing demand and supply scenarios around the world.

(c) **Providing Intelligence to procurement**: Advanced machine learning algorithms focus on past transactions, commodity prices, agriculture—based and industrial patterns, among others, to help companies understand where and when to source.

There will be more automation, more details, and more decisions to be taken, as procurement aims for digital transformation. For this, to handle large volumes of processes and the data procurement would require greater throughput and analytical capacity. That's where it comes to Artificial Intelligence (AI). AI will improve and enhance the capabilities of procurement by smart solutions that help work deliver transformative value [21].

(d) **Supplier risk management**: Risk management has to be a core component of supply chain management to ensure that the company is not subject to unnecessary threats, enforcement problems or negative publicity. For this, N-tier risk management helps organizations to model cost structures, trend performance data and visibility in the extended value chain to keep up to date with any supply disruptions and secure capacity.

(e) **Supply chain data management with intelligent automation and analytics**: In order to reliably capture supply chain transactions with high accuracy and low redundancy, end-to-end information management takes the form of some sort of data vaults. This will help companies in the supply chain to gain insights across supplier efficiency, diagnostics of the supply chain, market intelligence and risk management.

(f) **Supply chain simulation**: Modeling new supply chain approach is based on developments in the business/operational model, current and/or future supply/demand/logistic constraints. It helps to validate and identify the most cost-effective network in the value chain to achieve the level of service required.

The behaviour of a logistics network over time is demonstrated by a supply chain simulation. In a simulation model, the logical rules of a supply chain are interpreted and then executed over time, making the simulation dynamic. Production, for instance, starts when orders reach inventory below a threshold [22].

5 Post-COVID Time: When 'SCM Digitalisation' Will Spread Worldwide

(a) Digitization should break down barriers, so that the supply chain can become a completely integrated ecosystem that is fully transparent to all concerned parties, from suppliers of parts and components to distributors of those supplies and finished goods, and eventually to consumers who are seeking fulfilment. The "network" of digital supplies would offer a new degree of resilience and responsiveness, encouraging first-mover companies to beat the competition in an attempt to provide the most efficient and transparent service delivery to consumers.

From SCM point of view, as they are forced to confront their shortcomings and vulnerabilities, the current scenario is likely to intensify digital transformation initiatives for companies around the globe. Future global supply chain management business models will emerge as more important and relevant than ever and will play a key role in defining the strategy [23].

(b) There are many ways in which companies can create resilient supply chains in the post-COVID world, based on lessons that are reinforced and validated in the current global crisis. There is an urgent need in logistics and warehousing to reduce dependency on manual labour. This can be enabled for Industry 4.0 through core digital technology such as Internet-of-things (IOT), control towers, block-chain, demand forecasting enabled by rule-based and self-adjusting stock allocations, artificial intelligence/machine learning, autonomous devices such as drones and, Automated Guided Vehicle.

(c) Factories that can modularize production and shift/adapt lines due to changes in SCM demand will be the practice for the future. They will be supported by supply networks capable of interacting intelligently with each other, compounding their efficiency and agility. Businesses should pay great attention to making critical systems accessible on the cloud so that employees can access them remotely as they work from home.

(d) Safety will also be a crucial consideration and all planning initiatives will be at the core of supplier risk management. One of the few positive outcomes of the COVID-19 scenario has been introducing us to the possibilities of remote working across businesses, domains and industries, and this trend will lead to a renewed focus on environmentally sustainable operating principles if it is maintained in the post-COVID world.

(e) As we pass through the COVID-19 phase to the post-COVID world, the human element is the most important one that will emerge. Many major organisations, for example, include wording in their initial contracts requiring suppliers to engage in their supply-chain mapping activities on an annual basis. They should map their supply chains in detail, which includes identifying alternate sources of things, to ensure the same situation doesn't happen again.

6 Working Model

6.1 Algorithms or Models Used

(a) **Auto_Regressive model (AR_model)**: Based on past actions, an autoregressive (AR) model predicts future behaviour. It is used for forecasting when there is some correlation b/w values in a time series and the values which succeed and precede them.

With order of p AR model is shown in the Eq. 1.

$$y_t = c + \phi_1 y_{t-1} + \phi_2 y_{t-2} + \cdots + \phi_p y_{t-p} + \varepsilon_t \tag{1}$$

where ε_t is the white noise.

(b) **Moving _average model (MA model)**: A moving average model uses past forecast errors in a regression-like model rather than using past values of the forecast variable in a regression. The model is described by the Eq. 2.

$$y_t = c + \varepsilon_t + \theta_1 \varepsilon_{t-1} + \theta_2 \varepsilon_{t-2} + \cdots + \theta_q \varepsilon_{t-q} \tag{2}$$

where ε_t is the white noise.

(c) **Auto_Regressive_Moving_Average model (ARMA_model)**: In terms of two polynomials, one for autoregression (AR) and the second for moving average, this provides a parsimonious definition of a (weakly) stationary stochastic process (MA) shown in Eq. 3.

$$y_t = c + \phi_1 y_{t-1} + \cdots + \phi_p y_{t-p} + \theta_1 \varepsilon_{t-1} + \cdots + \theta_q \varepsilon_{t-q} + \varepsilon_t \tag{3}$$

(d) **Auto_Regressive_Integrated_Moving_Average model (ARIMA_model):**

ARIMA, is abbreviated form of 'Auto Regressive Integrated Moving Average' which is a class of models that 'explains' a given time series based on their own past values, i.e. their own lags and lagged forecast errors, such that future values can be predicted using the equation. With ARIMA models, it is possible to represent any 'non-seasonal' time series that exhibits patterns and is not a random white noise. An ARIMA model is characterized by 3 terms: d, q, p where,

1. p is the order of the Auto_Regressive term
2. q is the order of the Moving_Average term

3. d is the no. of differencing that is required for making the time_series stationary.

$$y'_t = c + \phi_1 y'_{t-1} + \cdots + \phi_p y'_{t-p} + \theta_1 \varepsilon_{t-1} + \cdots + \theta_q \varepsilon_{t-q} + \varepsilon_t \quad (4)$$

(e) Seasonal_Auto_Regressive_Integrated_Moving_Average with Exogenous variable model (SARIMA_model):

If you have seasonal trends in a time series, you need to apply seasonal terms and it will become SARIMA, short for 'Seasonal ARIMA'. Where we used lags = 15 in the equation Eq. 5

$$y_t = \beta 0 + \beta 1 y_{t-1} + \beta 2 y_{t-2} - 14 + \beta 3 y_{t-3} - 15 + \varepsilon_t \quad (5)$$

6.2 *Accuracy*

Root Mean Squared Error: By taking the square root of the mean squared error score. It can be transformed back into the original units of the predictions The mean squared error described as the average of the squared forecast error values This is called the RMSE or the root mean squared error.

$$\text{rmse} = \text{sqrt(mean\_squared\_error)}$$

This can be calculated using the mean_squared_error() scikit-learn and then taking square root using the sqrt() math function on it.

6.3 *Augmented Dickey Fuller Test Used for Checking If the Data is Stationary or Not*

We perform the dickey fuller test on the dataset to verify the data is stationary or not. The Statistical analyses make strong assumptions about the dataset called the Augmented Dickey-Fuller test.

It can only be used to indicate the degree to which a null hypothesis can or may not be rejected. For a given problem to be meaningful, the result must be interpreted.

However, they will provide a simple check and verifiable evidence that your time series is stationary or non-stationary.

A type of statistical test called a unit root test is the Augmented Dickey-Fuller test. The concept behind a unit root test is that it decides how strongly a trend defines a time series.

There are a range of unit root tests and one of the most commonly used may be the Augmented Dickey-Fuller. It uses an autoregressive model and optimises a criterion for knowledge over many different values of latency. The test's null hypothesis is that a unit root should represent the time series, that it is not stationary. The alternative hypothesis is the stationary time series.

(a) **Null Hypothesis (H0)**: If it is not rejected, it shows that the time series has a unit origin, which implies that it is non-stationary. It has a structure dependent on some time.
(b) **Alternate Hypothesis (H1)**: The null hypothesis is rejected; it means that there is no unit-root in the time series, implying that it is stationary. It does not have a structure depending on time.

P value > 0.05: If the null hypothesis (H0) is not dismissed, the data has a unit root and is non-stationary.

P value <= 0.05: Deny the null hypothesis (H0), there is no unit root for the data and it is stationary.

7 Proposed Methodology for Demand Forecasting in SCMs

(a) **Data Description**: dataset is taken from kaggle. It is the data of retail items supplied from 2014–2021. Having fields like date, price, stock and sale values. Where we will apply time series forecasting on sale with respect to time.
https://www.kaggle.com/tevecsystems/retail-sales-forecasting
(b) **Data cleaning and pre-processing**: In this step, the data entries in the dataset were removed which are having null or missing values and also all the date entries in data are converted in same format before data analysis. Analysis of time series provides a body of techniques for understanding a dataset better. The decomposition of a time series into 4 constituent components

1. Level. The baseline value for the series when it was a line which is straight.
2. Trend. The optional and sometimes linear increase or decrease of the series' behaviour over time.
3. Seasonality. Seasonality Optional continuation of patterns or behavior cycles over time.
4. Noise. Optional variability in the observations which cannot be clarified by the model.

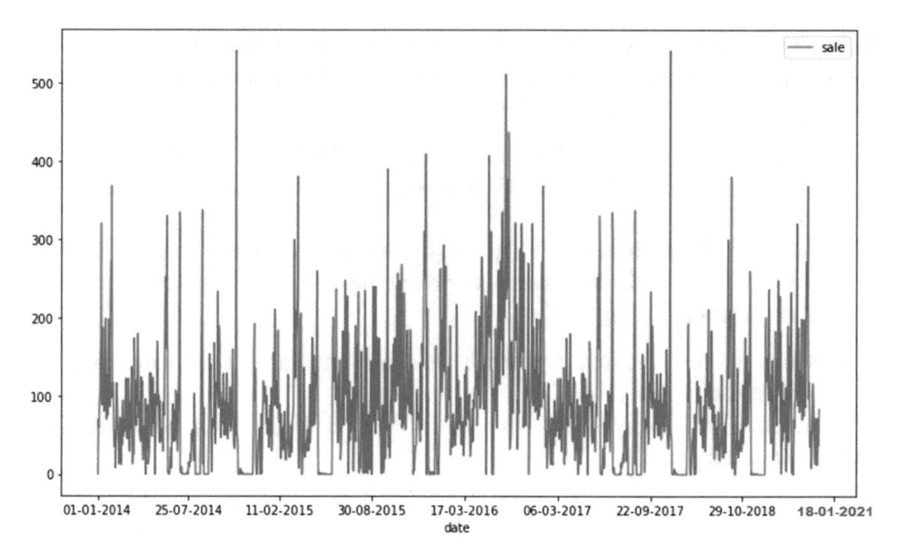

Fig. 1 Visualization of retail items sale versus time

Following are the steps of implementation:

Step 1: First Visualize the Time Series

Before constructing any form of time series model, it is important to analyse the trends. The information that we are interested in relates to some kind of trend in the series, seasonality or random behaviour. In the second section of this series, we covered this portion. The graph below shows the sale trend from year 2014–2021 (Fig. 1).

Step 2: Stationarize the Time Series

We can verify whether the series is stationary or not once we know the patterns, trends, cycles and seasonality. Dickey-Fuller is one of the most common tests to assess the same thing. In the first part of this article sequence, we've covered this test. There are 3 major used techniques to make a time series stationary are as follows:

1. Detrending: Here, we simply remove from the time series the trend variable. The equation of my time series, for example, is: x(t) = (mean + trend * t) + error. We will simply remove the part in the parentheses and create the remaining model.
2. Differencing: This is the approach widely used to eradicate non-stationarity. Here we try to model the terms' differences rather than the actual definition. For example,

$$x(t) - x(t-1) = ARMA(p, q)$$

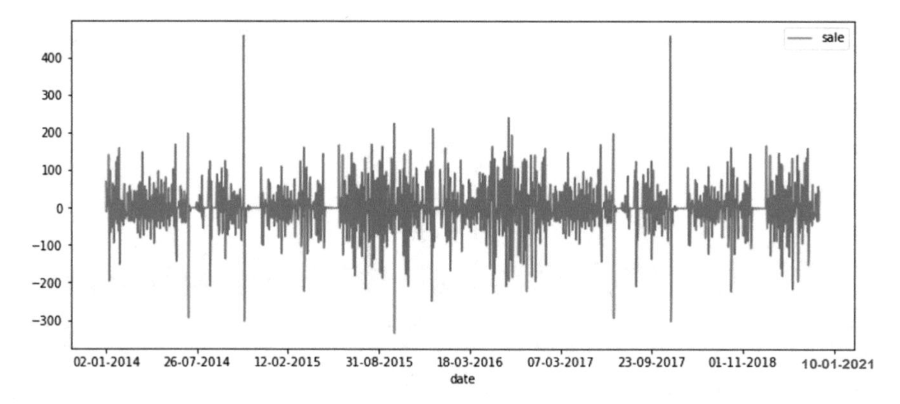

Fig. 2 Graph shows that data is stationary in nature

This distinction is referred to in AR(I)MA as the Integration component. We now have three parameters for p: AR d: I q: MA d:

3. Time series seasonality: You can easily integrate seasonality directly into the ARIMA model. In the applications part below, more on this has been discussed.

Data should be stationary for time series models. We have the data which is already stationary but If the data is not stationary than convert it from non-stationary to stationary.

The graph below shows the seasonality of data (Fig. 2).

Step 3: Search for optimal parameters

The AutoCorrelation function (ACF) and Partial Autocorrelation Function (PACF) maps the parameters p, d, q can be found. If both ACF and PACF progressively decrease, it means that we need to make the time series stationary and add a value to "d".

ACF: A number between −1 and 1 that represents a negative or positive correlation is the correlation coefficient of Pearson. No correlation implies a value of 0. Measure the correlation with previous time steps, called lags, for time series observations. Since the correlation of observations from the time series is calculated at previous times with values from the same series, this is called autocorrelation. The ACF is called a graph of the autocorrelation of a time series by lags (Fig. 3).

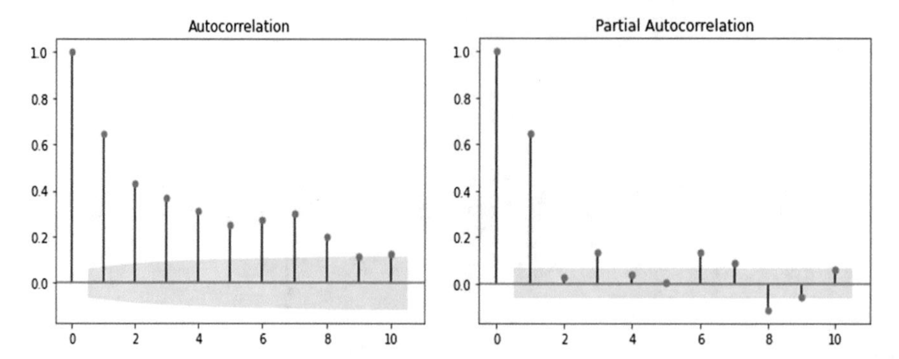

Fig. 3 Graph shows the ACF and PACF function

PACF: A partial autocorrelation is a description of the relationship between observations in a time series with observations at prior time steps with the relationships of intervening observations excluded. Both the direct correlation and indirect correlations make up the autocorrelation between an observation and an observation at a previous time point. The linear function of the correlation of the observation is these indirect correlations, with observations at the time steps interfering.

Step 4: Build an ARIMA Model:

We can now attempt to construct the ARIMA model with the parameters in hand. An estimated approximation may be the value found in the previous section and we need to explore more combinations of (p, d, q). Our option should be the one with the lowest BIC and AIC. We can also try some models with a part that is seasonal. Just in case, in ACF/PACF plots, we notice some seasonality.

1. Using ARIMA model without seasonality (Fig. 4)
2. Using ARIMA model with seasonality (Fig. 5)

Step 5: Checking error and accuracy of data using RMSE equation: after checking the seasonality now we can forecast data using time series models. The graph below shows the data predicted by the time series model and actual data with RMSE value of 6.9–7. We can also use MAPE, MAE, etc. in place of RMSE to check the accuracy of dataset. The graph below shows the actual data and the predicted data value generated by the model (Fig. 6).

Step 6: Make Forecasts

Once we have the final model of ARIMA, we are now able to make projections about future time points. If the model works well, we can also imagine the patterns to cross-validate (Fig. 7).

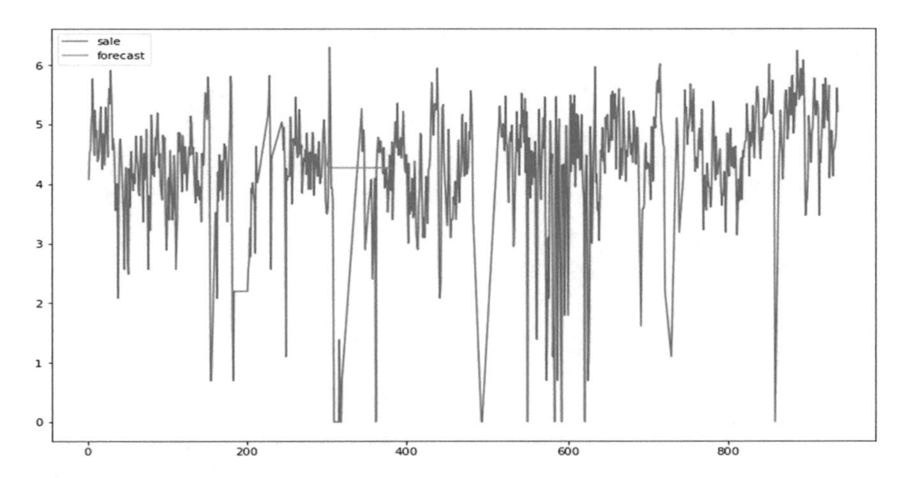

Fig. 4 ARIMA model without seasonality

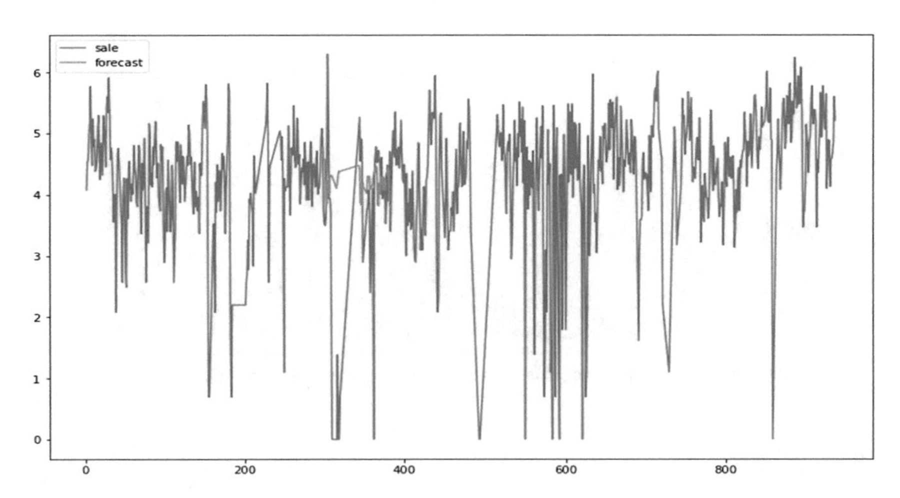

Fig. 5 ARIMA model with seasonality

8 Flow Chart

See Fig. 8.

9 Results

By applying the time series models on the data set we used ARMA MODEL to implement models like AR, MA, ARMA. ARIMA Model is used to implement ARIMA

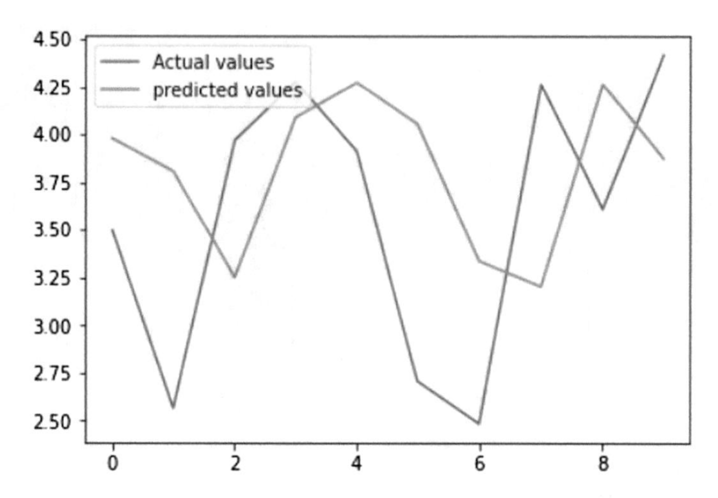

Fig. 6 Comparing actual and predicted value

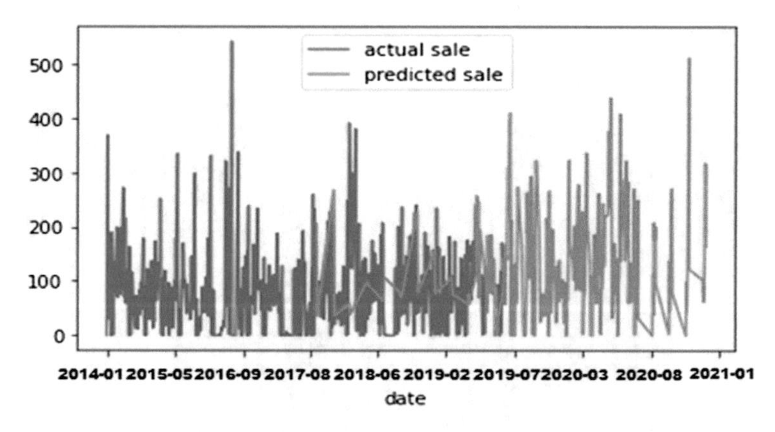

Fig. 7 Final sale predicted

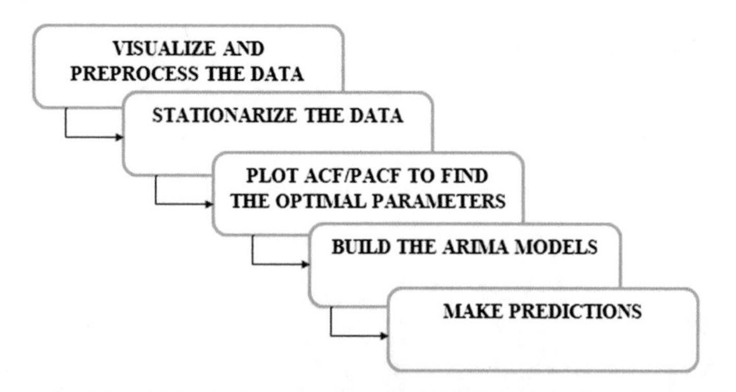

Fig. 8 Block diagram of the steps to implement time series analysis for SCM demand forecasting

AND SARIMAX model for SARIMA. To calculate the performance and accuracy of each model on the data set given. Then compare them using their AIC or BIC i.e. Akaike Information Criterion and Bayesian Information Criteria respectively. The values of either AIC or BIC can be taken to check the performance of the models. The model which is having smaller values of AIC/BIC is the best suited model for the dataset chosen which is used for time series forecasting. From the above summary of all the models which are AR, MA, ARMA, ARIMA, SARIMA having AIC and BIC values as shown in the below table:

s.no	Model name	AIC value	BIC value
1	AR model	2611.02	2625.87
2	MA model	2621.2	2641.01
3	ARMA model	2609.71	2634.47
4	ARIMA model	2606.68	2631.43
5	SARIMA model	3330.82	3356.69

Now plotting these AIC and BIC value to get a better view of all the models. The graph plots the values of the above table. Shown below (Fig. 9):

- From the above implementation of all the autoregressive models we get the RMSE values similar if we are implementing it on same dataset.
- So to compare these models we use the AIC or BIC values of all the models and compare them. The model having the smallest value of AIC or BIC is the most suitable model.

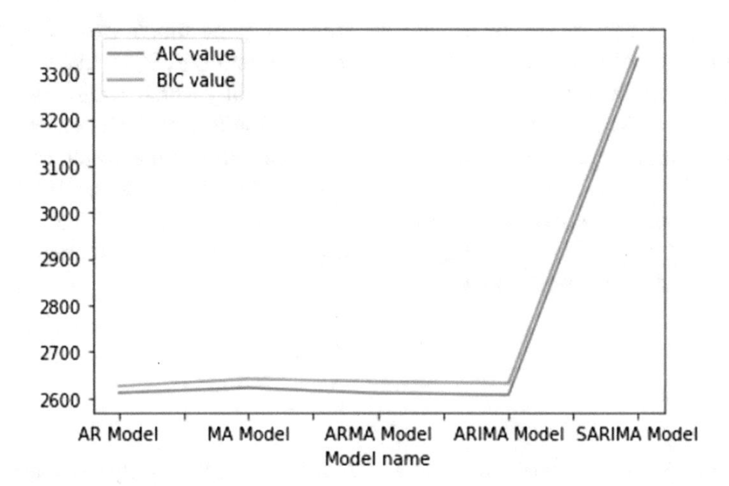

Fig. 9 AIC and BIC values of all the models

10 Conclusion

It is clearly shown in the above table according to AIC value of each model ARIMA model has the smallest value of AIC and that is the reason it is best suited for the given dataset for forecasting the demand in SCM but we can also use ARMA model because it has the nearby AIC value with the AR model.

On concluding that, by using the proposed model we get to know that ARIMA model is best suited out of all the Autoregressive models though we can also use ARMA model in place of ARIMA model. Retailers will know ahead of time what future demands will need to be met if the proposed approach is implemented. This model will help in COVID like situations, so that the companies get benefit of having lesser losses and manufacturer manufacture only according to the demands forecasted. By doing this there will be no wastage of materials. COVID-19 poses a number of serious and often unprecedented challenges for businesses that are cutting through the business environment from a strictly business perspective, including possible liquidity crunch, disruptions in the global supply chain, increased trade barriers, and shifting customer perceptions. However, digital technologies will play a critical enabling role in the post-COVID world in delivering changes across the breadth of companies, significantly enhanced user experiences, including more resilient supply chains, and intelligent optimised processes to deliver business results.

11 Future Work

More complex data set can be used as we were using the simple dataset. In future we can use the different data sets for demand forecasting in SCM like food items sale data, Medicines sale data etc. To select automatically which Time series model amongst all the auto Regressive models will be best suited for the given dataset we can use function already defined in R language i.e. auto_arima which selects the best model automatically from the autoregressive models. So we can also use R language whereas we have used Python language for our proposed work. We can also perform time series forecasting using deep learning approach whereas we have used the Machine learning approach.

References

1. Mollenkopf, D.A., Ozanne, L.K., Stolze, H.J.: A transformative supply chain response to COVID-19. J. Serv. Manag. (2020)
2. Sarkis, J.: Supply chain sustainability: learning from the COVID-19 pandemic. Int. J. Oper. Prod. Manag. (2020)
3. Montoya-Torres, J.R., Muñoz-Villamizar, A., Mejia-Argueta, C.: Mapping research in logistics and supply chain management during COVID-19 pandemic. Int. J. Logist. Res. Appl. 1–21 (2021)

4. Toorajipour, R., Sohrabpour, V., Nazarpour, A., Oghazi, P., Fischl, M.: Artificial intelligence in supply chain management: a systematic literature review. J. Bus. Res. **122**, 502–517 (2021)
5. Bäckstranda, J.E., Robert, S., Chend, R.C.: Purchasing process models. J. Purch. Supply Manag. **25**(5). https://doi.org/10.1016/j.pursup.2019.1005777
6. Kain, R., Verma, A.: Logistics management in supply chain—an overview. Mater. Today Proc. **5**(Issue 2, Part 1), 3811–3816. Pageno. Science Direct. https://doi.org/10.1016/j.matpr.2017.11.634
7. Vaidya, S.A., Ambad, P., Bhosle, S.: Industry 4.0—a glimpse. Proc. Manuf. **20**, 233–238 (2018). https://doi.org/10.1016/j.promfg.2018.02.034
8. Bode, C., Wagner, S.: Structural drivers of upstream supply chain complexity and the frequency of supply chain disruptions. J. Oper. Manag. **36**(1), 215–228 (2015). https://doi.org/10.1016/j.jom.2014.12.004
9. Sukati Abu, I., Hamid Rohaizat, B., Rosman, B., Yusuff, M.D.: The study of supply chain management strategy and practices on supply chain performance. Procedia Soc. Behav. Sci. **40**, 225–233 (2012). https://doi.org/10.1016/j.sbspro.2012.03.185
10. Schmidt, G.C., Wagner, S.: Blockchain and supply chain relations: a transaction cost theory perspective. J. Purch. Supply Manag. **25**(4) (2019). https://doi.org/10.1016/j.pursup.2019.100552
11. Fattah, J., Ezzine, L., Aman, Z.: Forecasting of demand using ARIMA model. Int. J. Eng. Bus. Manag. **10** (2018). https://doi.org/10.1177/1847979018808673
12. Chih, C., Tzu, P., Wen, C., Hung, J., et al.: Severe acute respiratory syndrome coronavirus 2 (SARS-CoV-2) and coronavirus disease-2019 (COVID-19): the epidemic and the challenges. Int. J. Antimicrob. Agents **55**(3) (2020). https://doi.org/10.1016/j.ijantimicag.2020.105924
13. Magal, P., Webb, G.: Predicting the number of reported and unreported cases for the COVID-19 epidemic in South Korea, Italy, France and Germany. BioRxiv and medRxiv (2020). https://doi.org/10.1101/2020.03.21.20040154
14. Helo, P., Hao, Y.: Artificial intelligence in operations management and supply chain management: an exploratory case study. Prod. Plan. Control 1–18 (2021)
15. Belhadi, A., Kamble, S., Fosso Wamba, S., Queiroz, M.M.: Building supply-chain resilience: an artificial intelligence-based technique and decision-making framework. Int. J. Prod. Res. 1–21 (2021)
16. Patil, M.: Challenges for supply chain management in today's global competitive environment. Eur. J. Bus. Manag. IISTE **7**(10). ISSN 2222-1905 (Paper) ISSN 2222–2839 (Online) (2015)
17. Rajah, N., Musa, H., Nipis, V., Kunjee Krishnan, P., et al.: Global supply chain management: challenges and solution. Int. J. Eng. Technol. 447–454 (2018)
18. Guan, D., Wang, D., et al.: Global supply-chain effects of COVID-19 control measures. Nature Hum. Behav. **4**, 577–587 (2020). https://www.nature.com/articles/s41562-020-0896-8
19. Raut, R.D., Gotmare, A.E., Narkhede, B., et al.: Enabling technologies for industry 4.0 manufacturing and supply chain: concepts, current status, and adoption challenges. IEEE Eng. Manag. Rev. **48**(2) (2020). https://doi.org/10.1109/EMR.2020.2987884
20. Olajide Ojo, O., Shah, S., Coutroubis, A., Torres Jiménez, M., et al.: Potential impact of industry 4.0 in sustainable food supply chain environment. In: IEEE International Conference on Technology Management, Operations and Decisions (ICTMOD). https://doi.org/10.1109/ITMC.2018.8691223
21. Zhao, Y., Shen, L., et al.: Application of time series auto regressive model in price forecast. In: International Conference on Business Management and Electronic Information. https://doi.org/10.1109/ICBMEI.2011.5921078
22. Priyamwadha, Wadhvani, R., et al.: Review on various models for time series forecasting. In: International Conference on Inventive Computing and Informatics (ICICI). https://doi.org/10.1109/ICICI.2017.8365383
23. Sundaresan, Y., Gupta, S., Sabeel, W.: Smart wearable prototype for visually impaired. Eng. Appl. Sci. **9**(6), 929–934 (2014)

Analysis of Supply Chain Management Data Using Machine Learning Algorithms

Khushi Arora, Prashant Abbi, and Praveen K. Gupta

Abstract Supply chain management is the regulation of the flow of goods and services that involves the movement and storage of raw materials, work-in-process stock inventory and finished goods. It also involves an end-to-end order fulfilment from point of procurement to point of consumption. This order fulfilment process in any supply chain management faces various uncertainties such as varied quality policies, multiple loading and unloading points, congestion along the route etc. Supply chain management system is the active management and regulation of supply chain activities in order to maximize the customer value and achieve a viable competitive advantage. It represents a conscious effort by the firms to develop and run these supply chains in the most efficient and effective ways feasible. Product sourcing, development, production, logistics, the information systems required to coordinate these activities all come under the purview of activities in the supply chain. The organizations that the supply chain comprises of are usually 'linked' together through physical and information flows.

Keywords Supply Chain Management · On-time delivery · Efficiency · Delivery · Correlation and mapping

Importance of efficient supply chain management systems

Supply chain management is a crucial part of most businesses. It is vital for a company's success and customer satisfaction. Efficiency if supply chain ensures:

K. Arora
Department of Computer Science and Engineering, R V College of Engineering, Bengaluru 560059, India

P. Abbi (✉)
Department of Information Science and Engineering, R V College of Engineering, Bengaluru 560059, India
e-mail: prashantabbi.is19@rvce.edu.in

P. K. Gupta
Department of Biotechnology, R V College of Engineering, Bengaluru 560059, India

K. Perumal et al. (eds.), *Innovative Supply Chain Management via Digitalization and Artificial Intelligence*, Studies in Systems, Decision and Control 424,
https://doi.org/10.1007/978-981-19-0240-6_8

- A boost in customer service since customers expect the accuracy in delivery, in terms of time, quantity, quality, location etc.
- Reduction in operational costs since an efficient supply chain would decrease purchasing and production costs thus leading to an overall decrease in total supply chain cost.
- Improved financial position since it increases profit leverage, decreases fixed assets and increases cash flow

The role of supply chain management in society is critical and not well known. Supply Chain Management can be used in a number of fields such as to support medical missions, conduct disaster relief operations, and handle other types of emergencies.

Machine Learning Algorithm: Classification using the Decision Tree

There has been a constantly increasing interest in classifying and predicting data from structured datasets. The algorithm we have used is Decision Tree algorithm which belongs to the family of supervised learning algorithms. Most supervised learning algorithms can be used to solve either regression or classification problems. Decision Tree algorithm can be used for solving both. The goal of using a Decision Tree is to create a training model that can be used to predict the class or value of the target variable by learning simple decision rules inferred from the training data. An added advantage is that we can represent any Boolean function on discrete attributes using the decision tree.

To predict a class label for a record, the process starts from the root of the tree in Decision Trees. The values of a root attribute are compared with the attributes of the record. Based on the comparison, the branch corresponding to that value is followed and the process repeats for the next node.

A heat map (correlation plot) was constructed for the features in the data. It has been observed that freight cost and insurance cost is directly correlated with timely delivery while, mode of fulfilment of the shipment are conversely correlated. Thus, we can infer that cost of the pharmaceuticals correlates to more timely delivery since they are most likely to be easily perishable. Scope for future research lies in finding more likely parameters to influence the time of delivery making supply chains more robust.

Proposed system:

This model predicts whether or not the delivery of the order placed will be on time. It is designed to keep a record of orders that are not delivered on time versus those that are. This output can be further used to find the efficiency of a given pharmaceutical company's delivery system and thus would be a reflection on the robustness of the company's supply chain management system. The proposed system would help customers track their deliveries too since most pharmaceutical orders tend to be critical in terms of timely delivery keeping factors such as type of medication, quantity ordered, whether or not the order is perishable etc into consideration.

1 Introduction

In the age of science and information technology, it is essential to leverage the use of existing resources to increase the efficiency of industrial processes. A supply chain comprises of all parties involved in the completion of a customer request, be it directly or indirectly.

Unnecessary faults and errors in supply chain management can prove to be detrimental to the profits of the companies by resulting in frequent revenue losses, increasing costs, and poor customer experiences. The uncertainty revolving around the delivery of the package triggers a series of business and operational issues. Typically, the delays are due to known causes, such as prices of commodities, multiple loading and unloading points, congestion along the route, quality policies, and process bottlenecks.

The pharmaceutical industry, in particular, contains multiple supply chains, with the end result of providing patients with medications on time. There are multiple challenges faced in the formation of an efficient chain including but not limited to drug-counterfeiting, price rigging, and cold shipping.

A key process that is vulnerable to inefficacies in supply chain management, especially for the pharmaceutical industry is the order-to-cash cycle. This cycle includes variables dealing with the inadequate inventory to fulfil demands, supply shortages and logistical uncertainties, huge backlog of orders, dynamic demand conditions, communication gaps among stakeholders, varied quality levels in the inventory and the return and delivery of products not ordered. These variables are dynamic and therefore differ accordingly, on a daily basis and hence are unlikely to be forecasted accurately using a set of pre-defined rules.

Machine learning and data analytics that allow real-time visualizations and action points can play a crucial role in understanding these challenges and predicting them well in advance. Having a layer of analytics on top of standard processes can allow businesses to take data-driven actions in real time (RM et al., 2020, Bhattachrya et al., 2020). Analysis of data that was previously hard to use is allowed by Machine Learning and can be used in the prediction models such as prediction of demand since a huge number of variables come into play, such as:

- Historical data
- Promotional activities throughout the market
- Recent and latest trends
- Existing competition in the market
- Market specifics
- Online presence.

Besides that, algorithms that are used in machine learning find new strategies or patterns that can be applied into the supply chain data with minimal manual involvement. This also accounts for provision of precise information and prediction that helps the organization. An added benefit of using machine learning algorithms is that large volumes of information can be processed very quickly, something that is not feasible through manual methods. In order to ensure consistent delivery, the processing of big data is important as supply chains generate large amounts of information at an international level.

Supply chains are able to attain enhanced accuracy in various departments of their business, including, but not limited to operations planning, workforce and logistics, by leveraging the robustness of machine learning technology. Hence, an algorithmic approach can drive operational efficiencies, reduce revenue losses from return orders, and improve the overall customer experience with on-time delivery.

This chapter includes:

- Significance of machine learning in Supply Chain Management
- Supply Chain Management of the Pharmaceutical Industry
- Literature Survey
- Proposed Methodology
- Conclusion and Future Scope.

2 Significance of Machine Learning in Supply Chain Management

With the help of machine learning, supply chain data patterns are quickly detected by algorithms that use significant variables for the success of the supply networks.

The timings and handovers involved with the movement of products in the supply chain can be accurately analyzed by machine learning techniques. This data can further be compared to existing benchmarks and with historical data in order to increase the efficiency of the supply chain by correctly identifying the bottlenecks. Additionally, the environment of the marketplace, prevalent seasonal trends, applicable promotions, sales and relevant historical analysis prove to be a good source of data for analysis. This data is then accordingly combined in order to maintain the demand and supply cycle by accurately predicting the demand for specific goods.

The presence of products in the correct place at the correct time ensure that supply chains are efficient. The upstream supply chain can be optimized by assessing the requirements of the customer with the help of machine learning. It fulfils the demand-supply cycle by providing goods aligned with the demands of the market.

Machine learning offers more precise catalogue management that helps to forecast the demand. In warehouse optimization, machine learning is used to help detect excesses and shortages of stock in the stores. Due to the ability to identify familiar schematics, audit storage and check the inventory every once in a while, in a more precise way, this is important in preventing sales losses. Quality is vital for good supply chain management as unnecessary rework and cost increases are caused by waste and faulty products. Machine learning can track how quality varies and suggest improvements over time. Apart from being applicable to products and materials, other areas such as shipping, suppliers and third-party quality can also be tracked.

One of the most difficult sections of Supply Chain Management is dealing with suppliers. The types of contracts, documentation and other areas that lead to the best results from suppliers can be analyzed by machine learning and used as a basis for future agreements and administration.

In order to enhance savings on the freight cost incurred and to provide a service that is more competitive, while also determining the impact on performance, companies and businesses also actively acquire Transportation Management Systems. Machine learning gives businesses the opportunity to access potentially insightful data and define ways by which the performance of the company can be enhanced.

There are several benefits of employing machine learning for Supply Chain Management:

- Machine learning techniques accurately optimize the flow of products thus allowing the organizations greater flexibility by not requiring them to maintain an 'over-stocked' inventory
- Waste reduction and significant quality improvement translates into a considerable reduction of costs
- Delivery of products can be ensured "just in time" in the marketplace for sale as a result of upstream optimization
- Simple and precise administrative practices enhance the supplier relationships
- Data Science, coupled with real time analytics enables the stakeholders and businesses to take more informed decisions by allowing them to continuously improve.

3 Supply Chain Management of the Pharmaceutical Industry

This chapter focusses on incorporating machine learning in the pharmaceutical industry's supply chain. A completely different set of goals, drivers, and constraints become dominant after a drug is launched. The supply chain of pharmaceuticals is the means by which prescription medicines are produced and delivered to patients. However, the supply chain network is complex, requiring a number of steps that need to be taken to ensure that medicines are made available to patients.

Multiple government agencies, hospitals, clinics, drug manufacturers, drug distributors, pharmacy chains, retailers, research organizations, and the FDA are involved in this supply chain. The same supply chain is responsible for the distribution of prescription medicines, over-the-counter (OTC) medicines, generics, as well as biologics with different handling needs and operational goals to compound matters further.

The involvement of numerous organizations including insurance companies and healthcare management organizations further add to the complexity of the chain, thus making the entire supply chain extremely difficult to maintain. Additionally, the rules and regulations governing the industry along with frequent mergers and acquisitions to enhance the research and development facilities contribute towards the exponential growth of unadministered methods of maintaining an optimum performance of the supply networks.

There are five well-defined steps in the pharmaceutical supply chain to ensure that providers and patients are given the required drugs as and when required.

Those five steps are:

- Pharmaceuticals originating from production sites.
- Shipped to wholesale distributors
- Stocked in retail, mail-order, and other pharmacy types
- Subject to price negotiations and processed by pharmacy benefit management companies through quality and management screens,
- Dispensed by pharmacies; eventually delivered to and received by patients.

Researchers note that the basic structure of the pharmaceutical supply chain has many variations, largely because of the constantly evolving players in the supply chain. Key players allow a smooth and efficient pharmaceutical supply chain network. Among these players are manufacturers, wholesale dealers, hospitals, and PBMs (Fig. 1).

The major issues that relate to the pharmaceutical supply chain are:

- Counterfeiting-related issues;
- Adverse drug reaction to patients;
- Increase in problems arising due to supply chain operations entities;
- Problems in the manufacturing processes such as the mixing of incorrect input raw materials, cross-contamination due to the manufacture in the same facility of more than one drug or inappropriate labelling of the finished product;

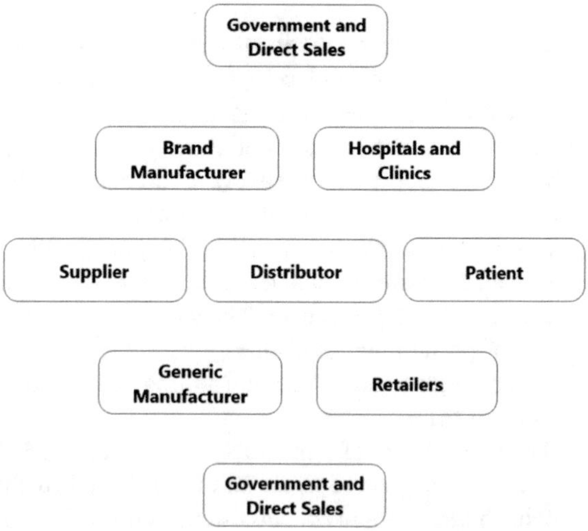

Fig. 1 Pharmaceutical supply chain

Government and Direct Sales
Brand Manufacturer
Hospitals and Clinics
Supplier
Distributor
Patient
Generic Manufacturer
Retailers
Government and Direct Sales

- Issues specific to retailers, including improper controls and handling of temperature;
- Transportation problems caused by mismanagement, improper temperature controls and the use of improper shipping mode;
- Storage problems, such as the use of improper temperature controls, improper warehouse handling and mixing of raw material products;
- Raw material suppliers' issues like improperly prepared raw material, raw material with high impurity levels and mislabeling of raw material shipments.

4 Literature Survey

Numerous articles and papers deal with the importance of incorporating machine learning techniques in supply chain management. However, despite being present for a long time, machine learning techniques haven't found widespread adoption in the supply chain industry until recently, with the advent of globalization. Lack of understanding in the developments in ML models might be a possible reason for the poor application of machine learning in SCM [2].

However, a few texts and models involving the use of Neural Networks, Support Vector Machines, Logistic Regression, Decision Trees and Extreme Machine Learning [5], were analyzed to indicate the penetration levels of machine learning techniques in the SCM sector.

The initial challenge prevalent in the field of supply chain management, involves finding new opportunities by establishing connections with suitable business partners such as customers or suppliers. This validates the facts regarding globalization and the fast development of technology. In order to simplify the method of finding new prospective business partners, Mori et al. [4] used enhanced machine learning techniques based on the profiles of companies and transaction relationships. To model existing relationships, a Support Vector Machine (SVM) was applied by using features inherent to the company, such as the number of employees and capital, as well as features defined by the interlinking of companies, such as supplier clients and common industrial categories.

Additionally, the need for an efficient forecasting model was established in the event when the actual demand information is unavailable and a method needs to be in place for establishing a strong chain [6]. This requirement was further validated by the fact that in supply chains, firms in the chain face what is known as variance amplification, that could be prevented by demand forecasting using machine learning techniques [3].

For adopting a model, Real Carbonneau et al. [6] was reviewed. This chapter worked on two separate datasets corresponding to the simulated supply chain and actual orders respectively. The findings proved that a regression or decision tree model could provide a higher forecasting accuracy, as opposed to the otherwise high performing neural networks and support vector machines. Further, in [1], it was asserted that random forests, a derivation of Decision Trees, used for both, classification and regression tasks would prove to be efficient when large databases are involved.

Hence, it was deemed fit to employ a model that could predict, whether a commodity or set of packages would be delivered on time or not. Such a model, upon further work and with additional information could prove to be useful in determining how various factors lead to a delay in the supply chain.

5 Proposed Methodology

5.1 Overview

The proposed methodology contains three phases: Data Pre-processing, Data Visualization and Model Training and Classification.

The data for this analysis was collected from an open- source website **Kaggle** in the form of a csv file named "supply chain shipment pricing data"**,** that contains readings recorded data of a few orders placed at various US Pharmaceutical companies in the period between May 2006 and December 2015 which amounts to a total of approximately 85000 readings. The **data mining** was done ensuring that all data pertaining to important details such as Product type, Shipment mode, cost of order, insurance price etc. were available and consistent.

In the **data pre-processing** phase, the string variables were grouped and accordingly mapped to integer variables so that they can be processed using the algorithms.

Data classification was done using the supervised classification algorithm—Decision Tree Algorithm.

Since the dataframe does have labels but isn't presented in a structured order, it needs to be pre-processed and presented leveraging the use of a supervised learning algorithm.

Unsupervised learning makes use of a dataframe that does not contain any labels. Accordingly, it attempts to find some organized structure within the available data.

Since our dataset has some basic structure, unsupervised learning algorithms will not be required (Fig. 2).

Supervised classification algorithms are in need of datasets with categorical dependent variables—'target' and sets of independent variables—'dependent variables' for determining a relation between them. The training procedure for classification works on the training set and maps the dependent variables to the required output variable i.e., target value, thereby creating a relation to correctly identify and group the recordings.

This training set contained 80% of the complete data while 20% was part of the test dataset (to determine the accuracy of the model). The training phase ends with generating the rules by quantifying the relationship of the output—target value, with the given input dependent variables.

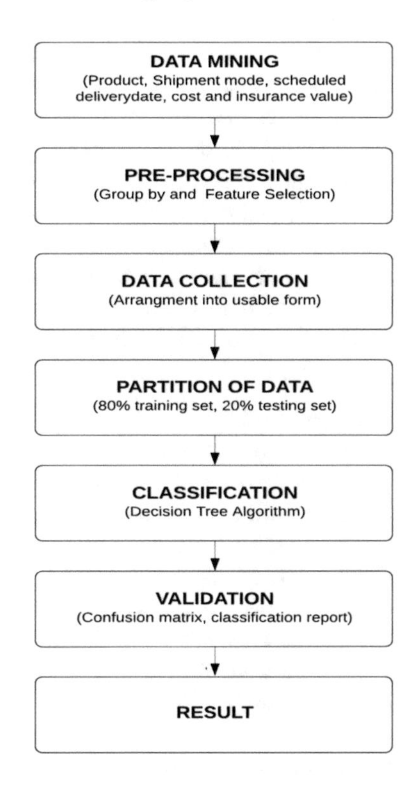

Fig. 2 Methodology flowchart

In the testing phase, the target value is predicted based on the generated rules. These predicted values are then compared to what the actual values were (for the 20% of the data) to determine the accuracy of the model. Key metrics such as the Pearson correlation coefficient, the confusion matrix and accuracy report are presented to give an accurate picture of the correctness of the model.

5.2 Data Collection—Obtaining the Dataset

The data used in this chapter is collected from an open-source repository, Kaggle, consisting of the details of orders placed in different pharmaceutical companies across various states in the United States of America between May 2006 and December 2015.

These were to be delivered to different countries throughout the world. Since these were overseas shipments, the mode of shipment varied from delivery to delivery. The Order ID, Project Code, Country to be shipped to, The office managing the order, Order fulfilment mode, Vendor INCOTerm, Shipment Mode, Date the price quote was sent to client, Date the price order was sent to Vendor, Date for when the delivery is scheduled, Date when the order is actually delivered, Date when the delivery was recorded, the group which the product belongs to, Sub Classification of the ordered medicine, Vendor that will be supplying the order, Description of the ordered item, Molecule/Test Type of the medicines in the order, Brand of the medicines in the order, Dosage of the medicines in the order, Dosage Form of the medicines in the order, Unit of Measure (Per Pack) of the medicines in the order, Total quantity of the order, Total value of the order, Pack Price of the medicines in the order, Unit Price of the medicines in the order, Manufacturing Site of the medicines in the order, Weight of the order, Freight Cost of the order and Order Insurance was recorded for orders. This resulted in recording 83929 readings in total.

The dataset contained the following dependent variables:

- Order ID
- Project Code
- Country to be shipped to
- The office managing the order
- Order fulfilment mode
- Vendor INCOTerm
- Shipment Mode
- Date the price quote was sent to client
- Date the price order was sent to Vendor
- Date for when the delivery is scheduled
- Date when the order is actually delivered
- Date when the delivery was recorded
- The group which the product belongs to
- Sub Classification of the ordered medicine
- Vendor that will be supplying the order
- Description of the ordered item
- Molecule/Test Type of the medicines in the order
- Brand of the medicines in the order
- Dosage of the medicines in the order
- Dosage Form of the medicines in the order
- Unit of Measure (Per Pack) of the medicines in the order
- Total quantity of the order
- Total value of the order
- Pack Price of the medicines in the order
- Unit Price of the medicines in the order
- Manufacturing Site of the medicines in the order
- Weight of the order
- Freight Cost of the order
- Order Insurance.

The given dependent variables are of the form—scheduled delivery date (DDMMYYYY), Order Value is US Dollars, Weight in Kilograms, dosage in milligrams, Pack Price, Unit Price, Freight Cost and Order Insurance in USD. dependent variables like Order ID, Unit of Measure (Per Pack), Line Item Quantity, Line Item Value are integer variables while Country to be shipped to, The office managing the order, Order fulfil mode, Vendor INCOTerm, Shipment Mode, Date the price quote was sent to client, Date the price order was sent to Vendor, The group which the product belongs to, Sub Classification of the ordered medicine, Vendor that will be supplying the order, Description of the ordered item, Molecule/Test Type of the medicines in the order, Brand of the medicines in the order, Dosage of the medicines in the order, Dosage Form of the medicines in the order, Manufacturing Site of the medicines in the order are all variables of type string.

This data is present in a csv file and needs to be converted to a suitable form for operating and analysing.

5.3 Data Pre-processing and Dependent Variable Selection

The first step to clean the data was to map all the variables of type string to an equivalent integer variable. Furthermore, an integral part of the pre-processing stage was to reduce the total number of observations and group them accordingly to streamline the data. Hence, the data was grouped by the **'Scheduled Delivery Date'** and **'Target'** dependent variables, aggregating the constituent entries into independent variables.

A heatmap with Pearson's correlation was plotted to check for redundant dependent variables, to visualize the relation between them and select the important ones. There are 2 types of correlations possible; positive and negative. A positive correlation indicates that if a dependent variable starts increasing, then the compared dependent variable increases as well, and if that dependent variable starts decreasing then the compared dependent variable also decreases accordingly. Both dependent variables vary according to a fixed pattern and they have a linear relationship. A negative correlation indicates that if the dependent variable is increasing then the compared dependent variable starts decreasing and vice-versa (Fig. 3).

From this heatmap, it is clear that the relevant dependent variables for the optimization of this analysis are Country to be shipped to, Order fulfilment mode, Vendor INCOTerm, Mode of shipment, Scheduled Delivery Date, Delivered to Client Date, Line-Item Value, Freight Cost and Line-Item Insurance since they have high correlation values. However, the other dependent variables were dropped since they had a low correlation with our independent variable, "Target" and wouldn't significantly influence the analysis.

This brought down the observations tremendously and the aggregation resulted in cleaner data. The final dataset contained the following dependent variables:

- **Country**—Country to be shipped to
- **Fulfil via**—Order fulfilment mode
- **Scheduled Delivery Date**—Date (YYYY-MM-DD) the observation was recorded;
- **Vendor INCOTerm**—The International Commercial Term of the Vendor
- **Shipment Mode**—The mode by which the order is being shipped
- **Scheduled Delivery Date**—The date when the order is estimated to be delivered

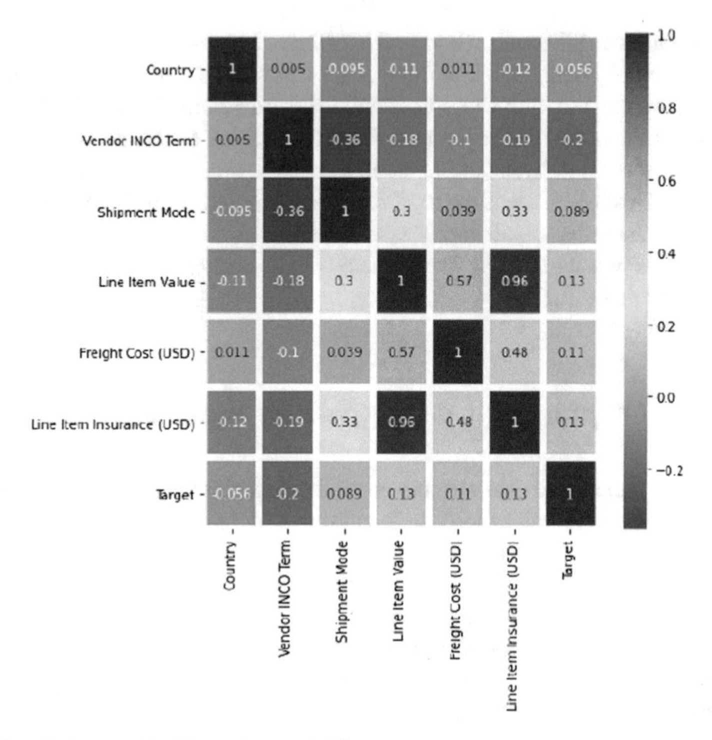

Fig. 3 Correlation matrix of dependent variables

- **Delivered to Client Date**—The date when the order was actually delivered
- **Line-Item Value**—Total value of the commodities in the order
- **Freight Cost (USD)**—Freight cost of the order in USD
- **Line-Item Insurance (USD)**—Insurance cost of the order in USD
- **Target**—Whether the delivery was on time or not (binary).

Hence, the final independent variable was identified as 'Target'.

5.4 Data Visualization

In this part, the processed data was plotted into various graphs and plots.

One bar graph named 'Top 10 Country Wise Count' showed which country had the greatest number of orders while another bar graph was the plot between Country and total order price. It was observed that even though South Africa had the greatest number of orders, Nigeria was the country that had the highest total pack price. This implies that the average order price of Nigeria was higher than South Africa's. Apart from this, a pie chart depicting the percentages of each shipment mode was also plotted (Fig. 4).

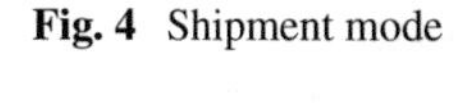
Fig. 4 Shipment mode

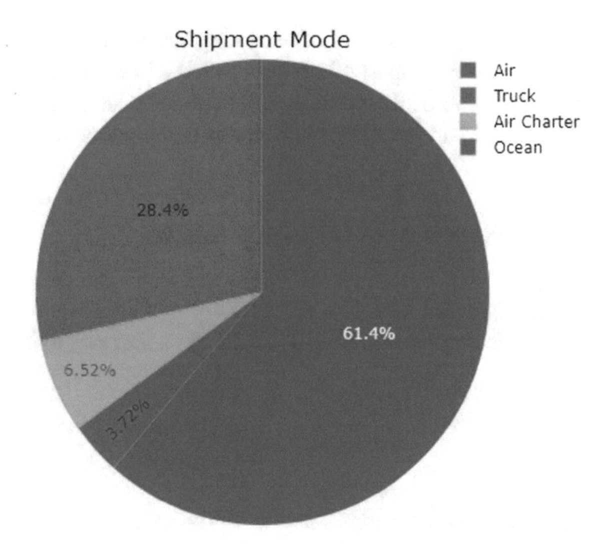

5.5 Algorithms Used in This Proposed Model

Classification process in machine learning comprises two steps. First being the learning step and the second being the prediction step. In the learning step, based on given training data, a model is developed. In the second step that is prediction, as the name suggests, the model is applied to predict the output for the given data.

The algorithm used is Decision Tree Algorithm for classification. It is a type of supervised learning. It is one of the popular classification algorithms. It is also relatively very easy to understand and implement. An added advantage to use the decision tree algorithm is that it comes handy while solving both classification as well as regression problems.

Using the Decision Tree algorithm, we have created a training model which will be then used to predict the value of the independent variable, i.e., 'Target' by learning simple decision rules deduced from the training data set.

In this algorithm, we assign a root and start from there to predict a class for a record. The value of the root element is compared with the value of the element of the record. On the basis of this comparison, the corresponding branch to that value is followed. The same is done with subsequent nodes.

Few important terms used in the Decision Tree Algorithm are:

- Root Node: The sample is depicted by this. The root node then is divided into two or more sets or branches.
- Splitting: The process of a node getting divided into two or more sub-nodes.
- Decision Node: A decision node is a sub-node that gets divided into further sub-nodes.
- Leaf Node: Also known as terminal nodes, these do not divide and do not have child nodes.

- Pruning: The process of removing sub-nodes of a decision node.
- Branch: A subpart of a tree is called a branch. It is also known as a sub-tree.
- Parent and Child Node: A parent node is the node that splits into sub-nodes while the child nodes are the sub-nodes that were created as a result of the splitting.

5.6 Results and Accuracy Report

A classification report comprises of

Precision: The capability of the classifier not to label a false sample as a true value.

$$precision = \frac{true\ positives}{true\ positives + false\ positives}$$

Recall can be defined as the ability of the classifier to find all samples that are true.

$$recall = \frac{true\ positives}{true\ positives + false\ negatives}$$

The F-beta score can be obtained by calculating the weighted harmonic mean of the recall and precision metrics. An F-beta score is between 1 and 0, with 0 indicating the worst and 1 indicating the best possible value.

Support is referred to as the number of occurrences of each class of the correct target values.

The measure of the accuracy of a Decision Tree Classifier is gauged by the construction of a confusion matrix. It is a summarized table used to define the overall performance and accuracy of a classification model for a test data frame for which the corresponding true values are known (Fig. 5).

Fig. 5 Classification report and confusion matrix

```
Classification report
              precision    recall  f1-score   support

           0       0.95      0.92      0.94       511
           1       0.54      0.64      0.58        70

    accuracy                           0.89       581
   macro avg       0.74      0.78      0.76       581
weighted avg       0.90      0.89      0.89       581

Confusion matrix
[[472  39]
 [ 25  45]]
```

6 Conclusion and Future Scope

This model has successfully been implemented to predict whether the delivery will be as the per the scheduled delivery date or not. It does not account for the exact number of days it will be delayed, which part in the supply chain is the reason for the delay, which parameter is to be optimized to make the supply chain more robust and efficient. This gives rise to a scope for future exploratory data analysis and research in the concerned domain. It is necessary to have timely deliveries and an efficient supply chain system. This accounts for high customer and client satisfaction, ensures least losses and agile functioning of the company's supply chain system. The proposed plan is to study the parameters causing delay in deliveries in a way that would help analyzing and improving their system. It would help them attain higher customer retentivity. Keeping in mind the future scope of the proposed plan, the model can be further updated to know the exact delay. This new information would help pin point the necessary parameters and by what extent it affects the delivery time and conditions.

References

1. Bhardwaj, R.: Applications of machine learning techniques in supply chain optimization (2018)
2. Bhattacharya. A., Kumar, S.: A systematic review of the research trends of machine learning in supply chain management. Trans. Res. Part C **38**, 73–84 (2014)
3. Feizabadi, J.: Machine learning demand forecasting and supply chain performance. Int. J. Log. Res. Appl. (2020). https://doi.org/10.1080/13675567.2020.1803246
4. Min, H.: Artificial intelligence in supply chain management: theory and applications. Int. J. Log. Res. Appl. **13**(1), 13–39, 84 (2009)
5. Ni, D.: A systematic review of the research trends of machine learning in supply chain management. Int. J. Mach. Learn. Cybern. (2020)
6. Carbonneau, R., Laframboise, K., Vahidov, R.: Application of machine learning techniques for supply chain demand forecasting. Department of Decision Sciences & MIS, John Molson School of Business (2011)

Supply Chain Based Demand Analysis of Different Deep Learning Methodologies for Effective Covid-19 Detection

K. S. Srujana, Sukruta N. Kashyap, G. Shrividhiya, C. Gururaj, and K. S. Induja

Abstract During Covid-19, many supply chains were disrupted. Supply chain resilience can be improved by developing business continuity capabilities using artificial intelligence (AI). This research examines how companies use artificial intelligence (AI) and looks at ways that AI can improve supply chain resilience by increasing visibility, reducing risk, and improving sourcing and distribution. Early detection of SARS-CoV-2 (2019-nCoV), which is caused by the lethal virus SARS-Cov-2 (Severe Acute Respiratory Syndrome Corona virus), has become critical especially as the epidemic spreads. X-rays and computed tomography scans are examples of medical imaging that can help in diagnosis. CT scans are preferable over RT-PCR tests because of their inaccuracy. In this era of fast technological growth, using artificial intelligence methodologies to construct models with a higher performance volume and better accuracy predictions is a huge step forward. Medical image analysis incorporating image processing and computer vision techniques were used to analyse the chest Radiographs and train the models. The accuracy and significant amount of data collection and prediction supply chain for efficient detection of COVID-19 utilising Artificial Intelligence techniques are described in this study. Models are built using data obtained by local CT scan centres. The data can be reviewed from time to time in coordination with CT scan centres. The application

K. S. Srujana · S. N. Kashyap · G. Shrividhiya · C. Gururaj (✉)
Department of Electronics and Telecommunication Engineering, BMS College of Engineering, Bengaluru, Visvesvaraya Technological University, Belagavi, India
e-mail: gururaj.tce@bmsce.ac.in

K. S. Srujana
e-mail: srujana.te17@bmsce.ac.in

S. N. Kashyap
e-mail: sukrutakashyap.te17@bmsce.ac.in

G. Shrividhiya
e-mail: shrividhiya.te17@bmsce.ac.in

K. S. Induja
Department of Computer Science (Artificial Intelligence), Andrew and Erna Viterbi School of Engineering, University of Southern California, Los Angeles, USA
e-mail: kanchisa@usc.edu

K. Perumal et al. (eds.), *Innovative Supply Chain Management via Digitalization and Artificial Intelligence*, Studies in Systems, Decision and Control 424,
https://doi.org/10.1007/978-981-19-0240-6_9

will provide accurate predictions so this has a significant impact on the tool's market worth. For the post-COVID-19 period, many firms are hastening the creation of management plans with supply chain transformation in mind. In this pandemic, but even so, the market will be even narrower, so without using a decentralised governance framework with an imbalanced structure among various markets, it really should be moved to a centralised management strategy that combines advantage of the existing strength of a blocked setup, with almost as much vicinity to the manufacturing countries and regions. In a stronger emphasis Supply Chain Management, value management, in value analysis, plays a significant role. Real-time raw data of chest CT scans from hospitals were considered and used it to train the model after pre-processing it. In a chest CT scan, multiple perspectives and organs are focused, but the work solely used the axial perspective of the lungs to prepare the dataset. Around 1900 photos of each COVID and Normal are included in the dataset. The data was pre-processed with a range filter for noise reduction, cropping, data augmentation, and other minor operations such as adjusting the image brightness and sharpness. Deep Learning algorithms are trained using this pre-processed data. VGG16, ResNet101, Inception v2, DesneNet169, and Mobile net are implementations of deep learning algorithms that is developed. The same dataset was used to train the above models, however because each model has a distinct architecture, the accuracy of the models varies slightly. The test dataset for all of the models includes 300 images in each class, and the findings demonstrate that DenseNet169 has the best accuracy among the models, while ResNet101 has the poorest. Furthermore, the medical image analysis of Covid-19 by several models aids in the selection of the most accurate model for COVID-19 predictions from CT scans. A windows application for the prediction of COVID has been developed where the user will upload the CT scan image and has an option to specify model and get the prediction. If tested positive, user can also view the infected area in the image. This application uses DenseNet169 as default model, in case user does not specify the model, as it performs the best. The user can view the COVID protocols and related queries from WHO website using a query button. The availability of CT scans and other commodities for this application varies as a result of the pandemic, which has an impact on the global supply chain and the market price of this tool.

Keywords Deep learning · AI · Computer vision · VGG-16 · ResNet101 · Inception V2 · Mobile net · DenseNet169 · Supply chain management · Transfer learning

1 Introduction

Companies have globalised supply chains in recent years to cut costs, but this has raised risk, due to Pandemic. Companies have begun to implement more risk mitigation methods, and this trend will continue to the next level. SARS-CoV-2 (2019-nCoV) is a pandemic viral disease caused by the Respiratory Syndrome. The current gold standard diagnostic technique for 2019-nCoV cases is to detect viral nucleic acid using reverse transcription polymerase chain reaction (RT-PCR). The sensitivity of some tests has been reduced, resulting in misleading negative results. The diagnostic methods are desperately needed. Furthermore, this RT-PCR molecular approach necessitates time-consuming operations in a highly regulated setting, with test results typically arriving in 4 h, restricting its widespread adoption. False negative RT-PCR test results, in particular, are a possible hazard to the public wellness. Numbers are cause for great concern, with about 44,748,380 people infected worldwide and 1,179,035 patients succumbing to death.

With such rapid COVID estimations, the supply chain GUI tackles unexpected complications that occur as a consequence of the pandemic. The market value of this equipment, as well as the availability of centres and radiologists, will have an impact on COVID's testing supply chain. The application's potency will help to increase its market value in the supply chain for COVID detection in advance.

To add to the regrettable truth, Brazil, India, Russia, South Africa and a lengthy list of the other 215 nations have lost the most to the North American continent, several of the top global innovators in medical services. In the United States, 9,120,751, with 5,933,212, recuperation cases and 233,130 deaths, were diagnosed overall in 2020. The risk of developing positive tested cases each day is growing rapidly, driving states and government bodies worldwide to lock down no compromise to prevent the disease.

The global response to prevent the spread of this disease was rapid and overwhelming, with the majority of afflicted countries barring their borders and limiting travel and transportation services. Individuals, states, nations and international corporations must follow formal recommendations issued by the World Health Organization and the Centres for Disease Control and Prevention (CDC) in order to entirely eliminate the disease and break the link that has led to this epidemic. The global WHO strategy for response COVID-19 includes 5 stages: (1) the rearming of the sanitary and social distance in all aspects of society, (2) control of asymptomatic individuals with a view to avoiding the dissemination of communities, (3) the reduction of spreading by Germane controls, (4) the supply of health system to control deaths and (5) the R&D of immunizations and the other regimes.

For illness diagnosis, medical imaging tools are essential. Image segmentation is an essential image processing step that separates areas of interest (ROIs) from the remainder of the image. Important properties such as tissue structure and texture can also be extracted via image segmentation. Image classification is achieved by extracting the import features from the images using a descriptor (e.g., SIFT and image moment) and then utilising classifiers such as SVM to employ these features in the classification task.

In comparison to handcrafted features, deep neural network-based approaches are more efficient at classifying images based on extracted features. Characteristics Based on the characteristics of ML, several attempts were made to categorise chest X-ray images into affected patient class or common case class using machine learning-based techniques. Neural networks has established a position in the artificial intelligence sector, which originally comes from the amount of hidden neurons it includes. For both computer vision-based image classification, regression difficulties, AI has generated encouraging results in the previous decade. Deeply CNN have increased the usage of images in the previous five years, driving them to new heights (Convolutional Neural Network, or CNNs).

In general, CNNs that attempted to imitate biological characteristics of humans on computers required picture or data pre-processing before feeding them to the network.

Each neural network is responsible for gathering data pertaining to the job. Although it is essential to imitate human activity and knowledge in neural nets, transfer learning is a separate field as it is utilised for transferring expertise from activity to activity. The pre-trained model will be freely available to all researchers to use the data they have collected. Models such as VGG16,13, VGG19,13 and ResNet101,14 were used to compare pre-trained systems available to the public.

2 Literature Survey

Based on initial survey of COVID-19 diagnosis studies, it appears that there is only a limited amount of chest radiography evidence for Corona infection detection and diagnosis [1]. Tools to image recognition Using chest CT imaging, COVID-19 can be detected. Traditional COVID19 detection techniques, such as PCR sets and Corona Detection tests (RT-PCR), have several disadvantages, including poor kit productivity, long turnaround times, and low pooled RT-PCR sensitivity (89%, for example).

When using a conventional PCR tester to identify this disease, false negatives are a real-time clinical issue. As a consequence, to address the difficulties that have arisen as a result of the circumstance. To identify COVID-19, scientists are developing towards using lung ultrasonography instead of PCR kits. Patients with COVID positive cases underwent analysis of chest X-ray imaging and chest CT scans [2].

Supply chain transformation has attained importance especially for the post-COVID-19 period, many firms are hastening the creation of management plans [3]. It is increasingly focused on automation, digitising the method, and the coordinating and reconciling of the wide range of ailments connected with the supply chain [4].

Ground-glass opacity can be seen on Diagnostic imaging of the lungs. According to this research, CT has a susceptibility of 98% for this illness, compared to 71% for RT-PCR. CT scans, on the other hand, are more successful for detecting COVID-19 because they offer a full complex 3d image of the body part and hence the irregularity's soundness (Fig. 1) [5–10].

With the aid of X-ray scans, they may be diagnosed more accurately. CT scan thoracic pictures of COVID positive people, on the other hand, are few. Thus, image classification combined with data augmentation has been shown to be a useful approach for identifying anomalies in a limited data of Corona patients' thoracic radiology [11].

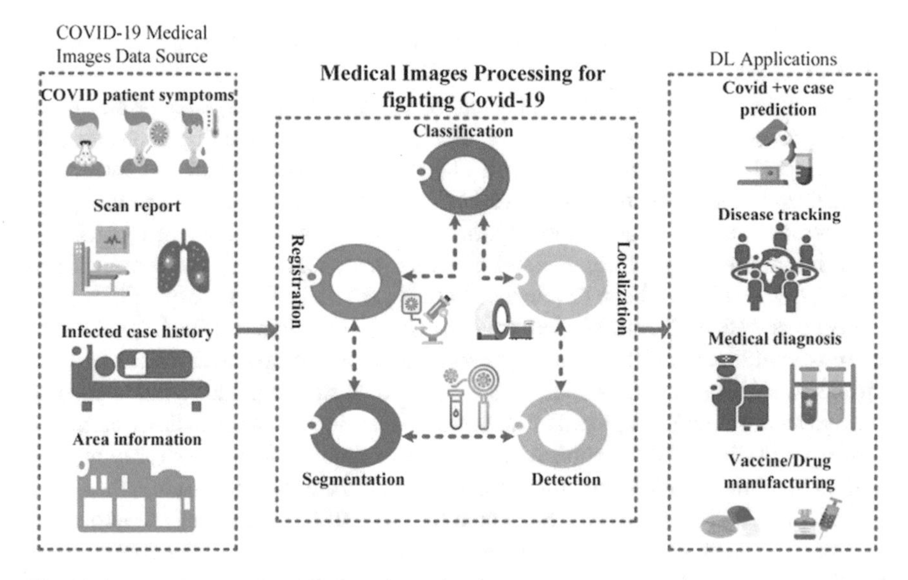

Fig. 1 Supply chain stakeholders for COVID diagnosis

In general, the X-ray evaluation workflow for COVID-19 includes the planning: (1) before the scan processing, (2) data acquisition, and (3) illness identification, using thoracic CT as an example. During the pre-scan preparation stage, a technician instructs and educates each participant on how to position themselves on the hospital beds in accordance with a standard. During the data acquisition step, CT pictures are recorded during a single inhalation and exhalation. The scan goes from the top of the lungs to the bottom. As from upper thoracic inlet up to the lower costophrenic angle, scans are done using optimum settings established by the radiographers depending on the patient's physique. CT images are reconstructed from raw data and sent to PACS i.e., Picture archiving and communication systems for viewing and assessment later [1].

X-ray penetrates as well as collects photographs of a certain segment in various perspectives for the CT scans. These photographs are stored to a machine, where they are subsequently processed to generate a new, overlapping-free image [12]. Interior structures' scale, composition, density, texture, and shape are all portrayed in these images, which aid physicians in better understanding them. As a result, CT scans are more often reliable analytical equipment compared to X-rays. This infection and many other cold-cough symptoms are undetectable on a lung CT or X-ray [13, 14].

Scanning devices are incredibly sophisticated machines and cleaning them for each individual patient is tough. Even if cleaning measures are done, there is a very significant danger of virus infiltrating onto the scan machine's surfaces. Swab samples have been demonstrated to be far more reliable than radiological approaches in aiding the identification of this disease. A thoracic Scan or X-ray is seen in many COVID-19 patients, although they fail the test thereafter [15]. In the fight against

COVID-19, artificial intelligence (AI), a relatively new tool emerging in the domain of Medical Sciences, has assisted [16].

In comparison to the traditional scanning system, which depends primarily on human labour, AI ensures the safe, efficient, and effective imaging solutions [17, 18]. A specialised imaging framework, lung and infection region categorization, clinical evaluation and course of treatment, as well as notable preclinical and practical research are among the most recent AI-powered applications in COVID-19. In addition, a number of commercial products that successfully incorporate AI to resist COVID-19 have been developed, showing the technology's capabilities [19]. On February 18, 2020, the Medical Imaging Computing Seminar (MICS),1 China's topmost coalition of radiation therapy researchers and start-up businesses, held its 1st virtual conference on COVID-19, with over ten thousand individuals in attendance.

Two recent significant pieces of work are an overview research on deep learning's future possibilities and a compendium of key therapeutic diagnostics for lymphatic system and intermittent respiratory disease detection and classification.

Microbleed identification, thyroid nodules recognition in CT images, automated pancreas segment, cell feature extraction and tracking, vertebral radiography score prediction, and multi-modal imagery categorization extensions are some of the topics covered [1, 11]. Computer vision and artificial intelligence, in particular, are significant facilitators for improving diagnosis [20] by rendering it easier to obtain findings that need to be addressed and assisting experts in their workflow. The state of the art basis for these technologies is swiftly proved by deep learning, leading to high accuracy. It also opened new borders with unprecedented rates of advancement in data analysis [21].

In order to better anticipate future barriers, problems and potential technological advances in the age of computer-vision learning, research activities are important to building large annotated image data sets [1, 21, 22]. The creation of new CNN variants and the availability of fast parallel solvers for contemporary GPUs has attracted a great deal of commercial interest from the DNNs. The principal strength of a CNN is its profound design [23], allowing it to recover a variety of distinctive characteristics at different protocol layers. A neural Network from the ground up is tough to create (or complete training). In the first place, CNNs need a huge set of training samples, that might be difficult to get in the area of medicine, where expertise annotations are costly and diseases (such lesion) limited [24].

The results of the Literature Survey highlights the emergence of five critical areas where AI can contribute to enhanced supply chain resilience; (1) transparency, (2) ensuring last-mile delivery, (3) offering personalized solutions to both upstream and downstream supply chain stakeholders, (4) minimizing the impact of disruption and (5) facilitating an agile procurement strategy.

3 Dataset

The data for this study comes from the GitHub repository of the Johns Hopkins University Center for Systems Science and Engineering. The university originally made this repository available via the visual dashboard of 2019 Novel Coronavirus. The training dataset was segmented using AI-based technologies designed specifically for radiological image segmentation (Fig. 2).

Scanning machines are incredibly sophisticated machines, and cleaning them for each individual patient is tough. Even if cleaning measures are done, there is a very significant danger of virus leaking onto the scan machine's surfaces. For COVID-19 detection and diagnosis, swab samples have been demonstrated to be more reliable compared to image processing-based approaches. Many COVID-19 patients have a normal chest CT or X-ray yet test positive for the virus later on.

For real time images, dataset was obtained from Guru CT Scan centre, Mukkidipeta, Hindupur—515201.

There are many views of lung images in a CT scan i.e., Axial view, Normal lung view, Sagittal view [25]. Radiographer has suggested to use Normal lung view to analyse the CT scan better. These images were classified into Normal class and Covid class with the help of radiographer. The ground truth of the scans is taken and after predicting these images using the models developed, we can find the accuracy and other evaluation parameters of the model (Fig. 3).

Fig. 2 Training dataset

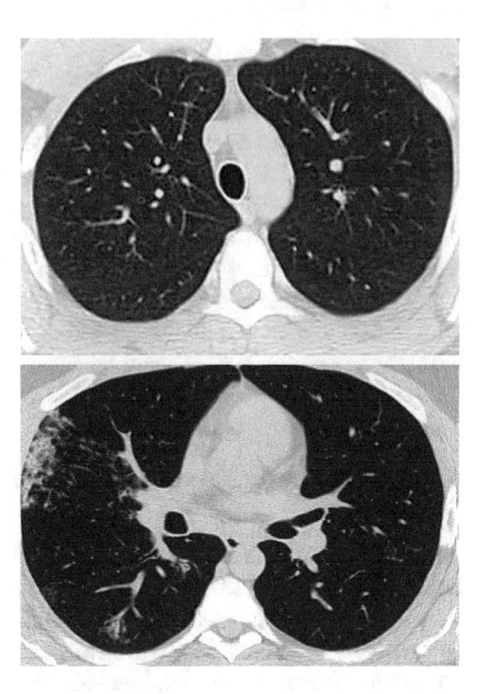

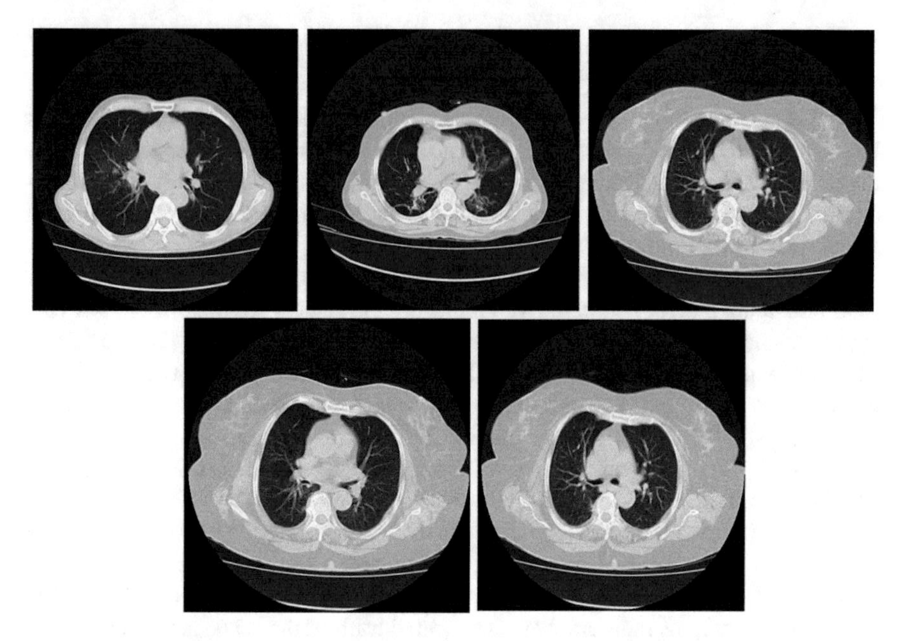

Fig. 3 Real time dataset

4 Methodology

The real time detection of covid-19 via CT scans of patients can be done accurately if only the models are trained properly. For that we put our obtained dataset into a series of procedures as shown below in the figure. An overall training flowchart can be seen, after which testing with real time information happens (Fig. 4).

With data processing and augmentation, over fitting of the model can be controlled and suppressed. Convolutional neural network forms the fundamental approach from which transfer learning can be utilized to process other CNN based algorithms like VGG-16 and further other models. The model which performs accurately when tested with real-time data will be considered the best among other models in terms of all evaluation parameters which will be discussed in further sections.

4.1 Preliminary Processing of Data

Processing of data before training is a crucial part of image processing techniques. AI algorithms including ML and Deep Learning work effectively when pre-processed data is fed as the input. Pre-processing is a technique to modify the given data so that using modified data will give a significant change in the efficiency of the algorithm. As a matter of fact, Machine Learning algorithms needs to be first trained

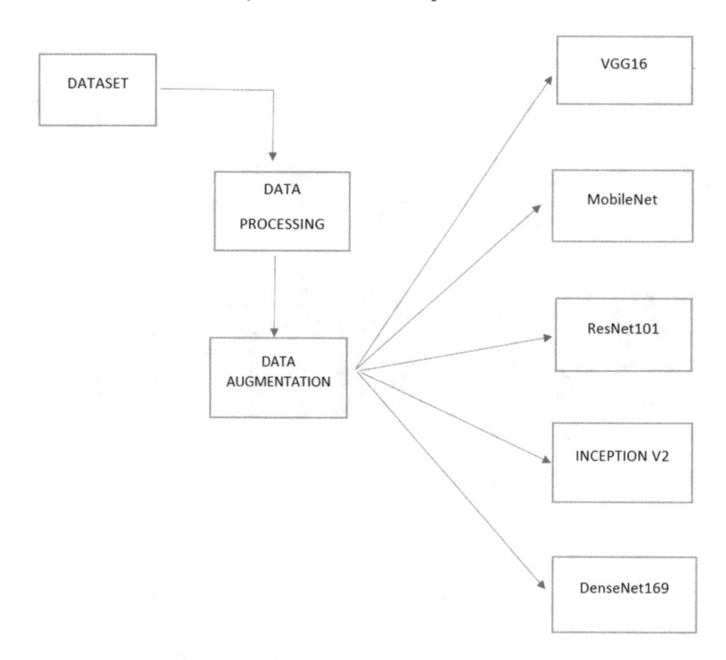

Fig. 4 Showing overall framework of operations for training

at the information available after which validated and tested, earlier than it could be deployed to address real-global records.

Here the images obtained from the CT scans are of pang format. These images when played as a set of frames with delay is video. Now using these images, we make NIFTI images which are basically like the videos which aids in understanding the organ for our analysis [26, 27].

These images will be loaded and mask will be applied to get the ground truth of the image array. The mask helps in identifying the infected part of the CT scan in detail. Pre-processed image will give an identifiable part than raw image. Hence it is important to adjust the brightness and sharpness of the image after cropping it. The masks for few sample images are shown below (Fig. 5).

4.2 *Need of Transfer Learning*

Pre-trained models' availability has reduced time and thus can be used for dataset of any particular requirement. Rather than the traditional machine learning where all the models irrespective of their repetitive use had to be trained every single time. So, this reuse of pre-trained models is known as transfer learning. Not only can the model, other information regarding the datasets also be shared. Nevertheless, given

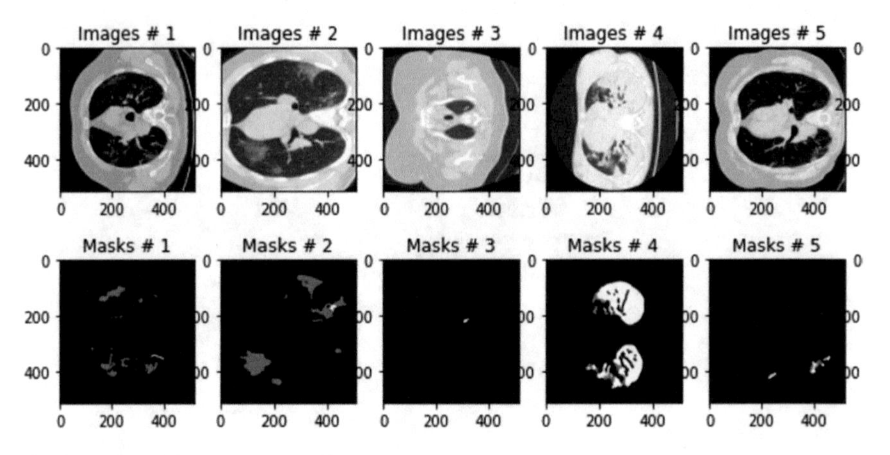

Fig. 5 Masks of the pre-processed images

the huge resources required for the education of profound learning models and enormous complicated data sources, transferring learning is common in programming improvement [3, 4, 20, 28–30].

Learning process is isolated, based only on specific objectives, data sets and the learning of separate, individual models. No intelligence may be passed from one design to another. Transfer learning lets you to leverage knowledge from prior trained models (features, weights, etc.) to train newer models and even address difficulties such as lower data for a new task (Fig. 6).

The model created utilising transmission learning has previously been familiar with its work and so no additional comprehensive coding is necessary. By transferring, you may use experience of formerly trained models for new ones and even solve obstacles like fewer data for the new job.

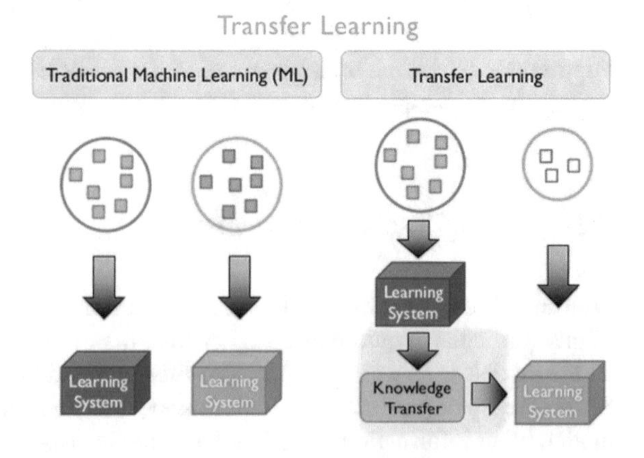

Fig. 6 Shows traditional ML versus transfer learning

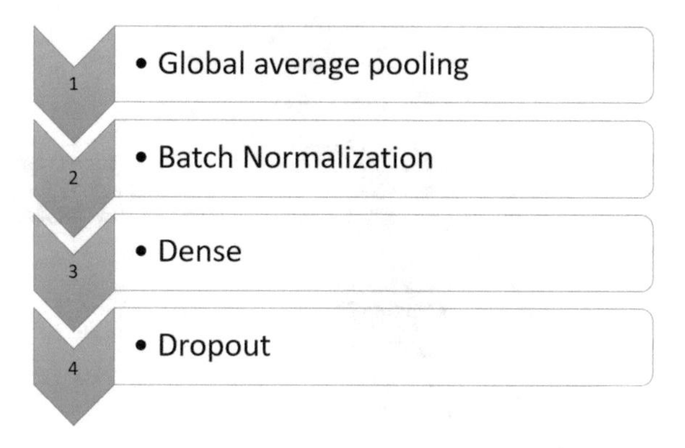

Fig. 7 Padded layers

The best model will be taken into account for each algorithm at the 50th epoch, and weights utilised for data testing. Based on the validation categorical precision, the best model is considered.

Padded layers (Fig. 7):

These layers were padded in certain order for each model to enhance the model performance with respect to the dataset. These padded layers will be added to the base model which was obtained by ImageNet weights using Transfer learning. Final layer to be padded will be the Dense Layer with SoftMax as activation function for all the models as it is preferable for Binary classification.

4.3 Global Average Pooling

This layer is used to replace fully connected layer where it creates a feature map by taking average of each convolution channel. This reduces the total number of parameters in the model minimizing overfitting. A tensor will be of h * w * d dimensions, when this average pooling is applied to each channel it takes the average of h * w channel and gives one single value which will reduce the dimension of the tensor to 1 * 1 * d. This extreme dimensionality reduction will reduce the overfitting of the model (Fig. 8).

Batch Normalization:

This regularisation approach, as the name indicates, normalises the output of the preceding layer by computing the batch average and deducting and dividing it by the batch standard deviation. A batch's characteristics might range from one extreme to the other. As a result, normalising the batch would benefit the model train. This decreases the having to rely of gradients on input value levels. Deeper networks can further decrease the effect of overfitting by using batch normalisation to reduce internal covariant shift and turbulence in layer activation distribution.

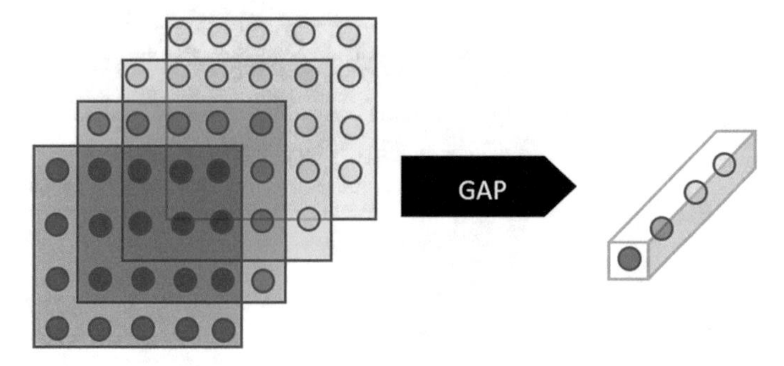

Fig. 8 Global average pooling

Dense Layers:

In Deep Learning, it is primarily a deeply/fully linked layer. So, all layer's neurons receive information from the previous layers, making the process completely linked. It will comprehend each component, each feature of the input when it is fully linked. This degree of scrutiny will help the model anticipate accurately. Because they are present between input and output, all dense layers are hidden layers. This dense layer calculates weights and passes the output to the next layer using activation functions for each activation unit, i.e., neurons.

For each activation cell, weights will be computed. The base model will also include particular weights of its dense layers, which we will use and mount the weights generated by the padded dense layers using transfer learning to improve our model (Fig. 9).

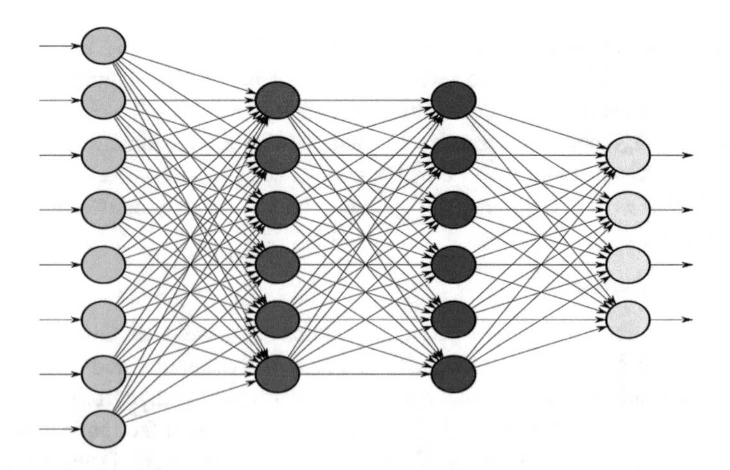

Fig. 9 Dense layer

4.4 Dropout

The key to reducing overfitting models is to use dropout layers. Dropout is a regularisation technique for deeper networks. Several units will be dropped in Dense layers at random. The tensor's input and output connections are also lost when it is dropped. If significant units are eliminated and most of the remaining units include information that is similar, this might lead to additional overfitting of the model. This replication must be addressed.

The dropout rate can be set, and the remaining tensor weights should be synchronised with normalised values by multiplying it by 1/(1 − dropout rate). Dropout simulates sparse activation from a specific node, which, as an unanticipated side effect, enables the network to learn a sparse representation. As a result, it might be utilized instead of activity normalisation in autoencoder models to boost sparse representations.

4.5 VGG16

It is one of the Convolutional neural network architectures which contains 16 layers (representing 16 in VGG16) in which 13 are convolution layers with 5 max pooling layers for down-sampling the images in between the convolution layers and in the end with 3 Densely connected layers. Its layers are as follows

(1) Input Layer: Images of 3 different channels which are resized to 224 × 224 are taken as an input.
(2) Convolution Layer: Images given as input will go through many convolution filters with kernel size 3 × 3 with given stride length say 1, 2 or 3.
(3) VGG16 is an architecture of convolution neural network (CNN).
(4) The activation function called ReLu is present in the hidden layers. At the output SoftMax is applied.

The working of this model requires the images to be resized to 224 × 224 here.

A succession of convoluted layers (conv) with a very tiny sensory field analyse the image: 33 (the smallest size capable of capturing left–right, up/down, and centre notions). For one arrangement, 11 convolutional filterings are being used, which may be viewed as a linear shift in input channels. The convolution step is set to one pixel, and the spatial padding of the convolution layer input is set to 1 pixel, which preserves space resolution following convolution, for 33 convolution layers. Spatial pooling is done via five max-pooling layers that follow part of the conv. layers (not all the conv. layers are followed by max-pooling). Stride 2 × 2 across a 22 pixel window with a stride of 2 is used for max-pooling.

The very first 2 have 4096 channels apiece, while the other has an ILSVRC 1000-way classification and hence 1000 channel classification. Three completely integrated (FC) layers are added following a pile of convolutionary (varied depth in

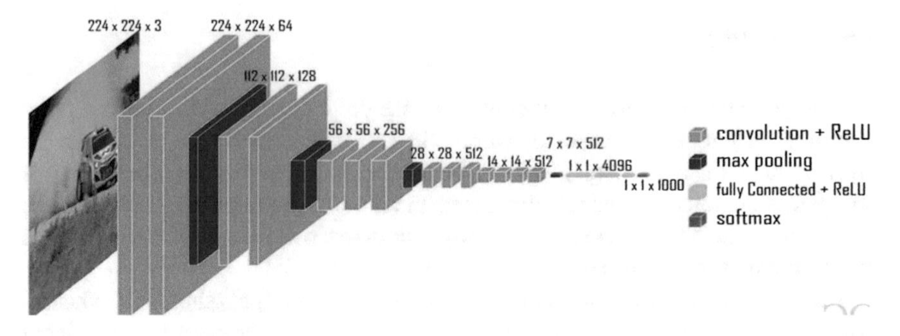

Fig. 10 VGG16 (*Source* https://neurohive.io/en/popular-networks/vgg16/)

various designs) (one for each class). The very last layer is the layer of the soft Max. All networks have the completely linked levels in the same configuration (Fig. 10).

4.6 ResNet101

Another exemplary deep network is ResNet. It's a robust backbone model that's employed in a variety of computer vision jobs. It solves the gradient issue (since the value is back-propagated to previous layers, many loops of calculation of that can result in a very small gradient). As a result, as the network becomes more complex, its performance becomes saturated or even degrades rapidly) as shown in deep neural networks. ResNet makes advantage of a unique feature called skip connection.

The connection skip is shown in the following diagram. The figure on the left displays turbocharged layers on each other. To the right, we keep on stacking convolutional layer as before, but now we also add the original input to the output of the convolutional block. The link is called a skip connection (Fig. 11).

Two types of shortcut modules are available in the ResNet implementation. The first is a block of identity with a bypass level without convolution. The input and output in this case have the same size. The other is the convolution block that has an adjustable layer shortcut. The inputs proportions in this situation are less than those in the output. To start and complete the network has 101 layers. This approach is called a bottleneck which, while preserving networking speed, reduces the amount of

Fig. 11 Skip connection

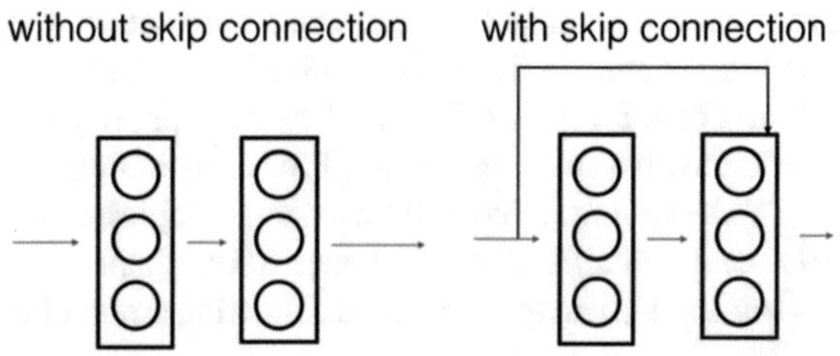

without skip connection

with skip connection

layer name	output size	18-layer	34-layer	50-layer	101-layer	152-layer
conv1	112×112	7×7, 64, stride 2				
conv2_x	56×56	3×3 max pool, stride 2				
		[3×3, 64; 3×3, 64] ×2	[3×3, 64; 3×3, 64] ×3	[1×1, 64; 3×3, 64; 1×1, 256] ×3	[1×1, 64; 3×3, 64; 1×1, 256] ×3	[1×1, 64; 3×3, 64; 1×1, 256] ×3
conv3_x	28×28	[3×3, 128; 3×3, 128] ×2	[3×3, 128; 3×3, 128] ×4	[1×1, 128; 3×3, 128; 1×1, 512] ×4	[1×1, 128; 3×3, 128; 1×1, 512] ×4	[1×1, 128; 3×3, 128; 1×1, 512] ×8
conv4_x	14×14	[3×3, 256; 3×3, 256] ×2	[3×3, 256; 3×3, 256] ×6	[1×1, 256; 3×3, 256; 1×1, 1024] ×6	[1×1, 256; 3×3, 256; 1×1, 1024] ×23	[1×1, 256; 3×3, 256; 1×1, 1024] ×36
conv5_x	7×7	[3×3, 512; 3×3, 512] ×2	[3×3, 512; 3×3, 512] ×3	[1×1, 512; 3×3, 512; 1×1, 2048] ×3	[1×1, 512; 3×3, 512; 1×1, 2048] ×3	[1×1, 512; 3×3, 512; 1×1, 2048] ×3
	1×1	average pool, 1000-d fc, softmax				
FLOPs		1.8×10^9	3.6×10^9	3.8×10^9	7.6×10^9	11.3×10^9

Fig. 12 ResNet variants (*Source* https://neurohive.io/en/popular-networks/resnet/)

variables. In the situation that the network is thicker, ResNet101 exceeds significantly. Shown below there are different types of ResNet architectures (Figs. 12 and 13).

4.7 *MobileNet*

This is a convolutional neural network architecture which uses an inverted residual block. The residual connections between the bottleneck layers are included in this structure, which is the complete opposite of the blocks described in ResnNet101.

In the intermediate expansion layer, Mobile net utilises lightweight depth wise separable convolutions to filtering out the features. In order to retain representational power, non-linearities in the thin layers also are eliminated (Fig. 14).

The expansion layer contains a depth wise layer, batch normalization layer and ReLu as its activation function. A total of 16 residual blocks will be used to form the mobile net architecture. The skip connections are also used between the bottleneck layers where the residual connection is present (Fig. 15).

This version uses drastically lower parameters compared to the previous version. The figure below gives us the details about layers and the dimensions of the images at each layer. The depth wise separable convolution layer was introduced to reduce the cost of complexity in the original mobile net version. This was mainly intended to perform well on mobile devices hence the name mobile net (Fig. 16).

4.8 *Inception v2*

This is one convolutional neural network architecture which is implemented for image recognition problems. This model is the modified versions of other two models

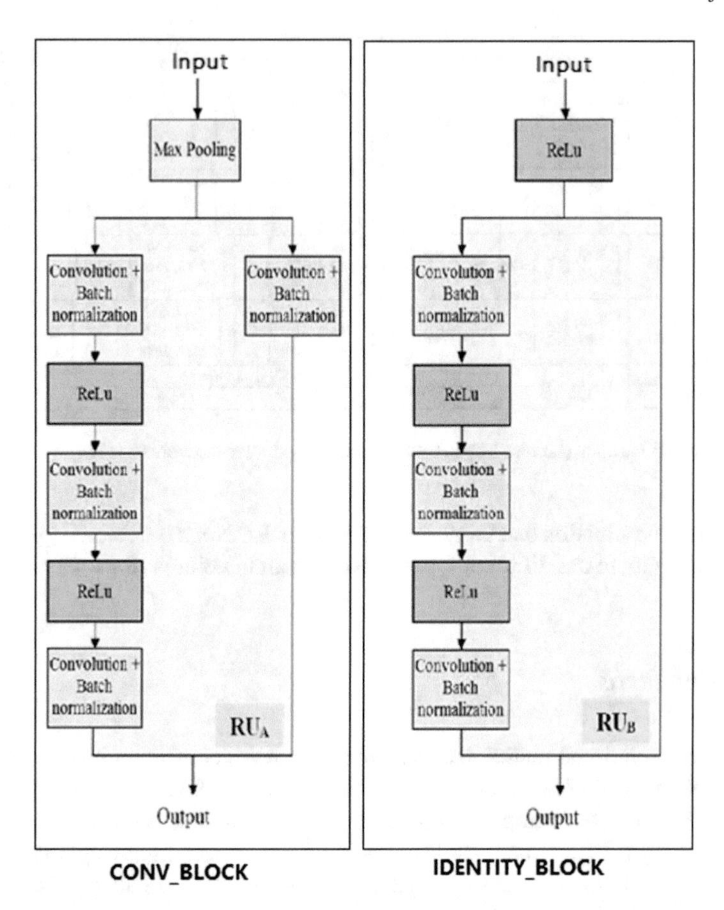

Fig. 13 ResNet101

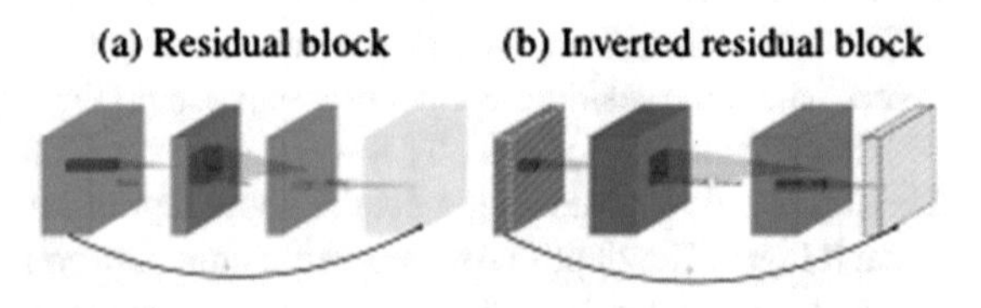

Fig. 14 Inverted residual block (*Source* https://towardsdatascience.com/mobilenetv2-inverted-residuals-and-linear-bottlenecks-8a4362f4ffd5

from the inception family. This model has best computational efficiency compared to the other models we have implemented. This computational efficiency is offered due to the presence of many parallel layers in one inception module. The dimension reduction happening in between the inception modules also helps in better computational cost compared to other models (Fig. 17).

Label smoothing and the use of an auxiliary classifier to transfer label information across the network, including the batch normalisation layers, are included in the upgraded version, as well as the addition of factorised convolutions with a kernel

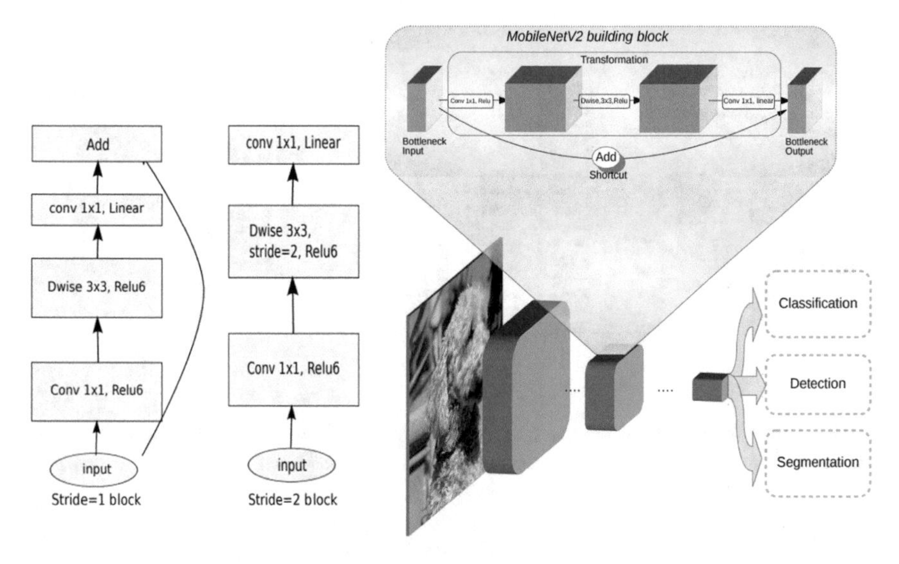

Fig. 15 Mobilenet architecture (*Source* https://paperswithcode.com/method/mobilenetv2)

Input	Operator	t	c	n	s
$224^2 \times 3$	conv2d	-	32	1	2
$112^2 \times 32$	bottleneck	1	16	1	1
$112^2 \times 16$	bottleneck	6	24	2	2
$56^2 \times 24$	bottleneck	6	32	3	2
$28^2 \times 32$	bottleneck	6	64	4	2
$14^2 \times 64$	bottleneck	6	96	3	1
$14^2 \times 96$	bottleneck	6	160	3	2
$7^2 \times 160$	bottleneck	6	320	1	1
$7^2 \times 320$	conv2d 1x1	-	1280	1	1
$7^2 \times 1280$	avgpool 7x7	-	-	1	-
$1 \times 1 \times 1280$	conv2d 1x1	-	k	-	

Fig. 16 Layers in mobile net (*Source* https://towardsdatascience.com/mobilenetv2-inverted-residuals-and-linear-bottlenecks-8a4362f4ffd5).

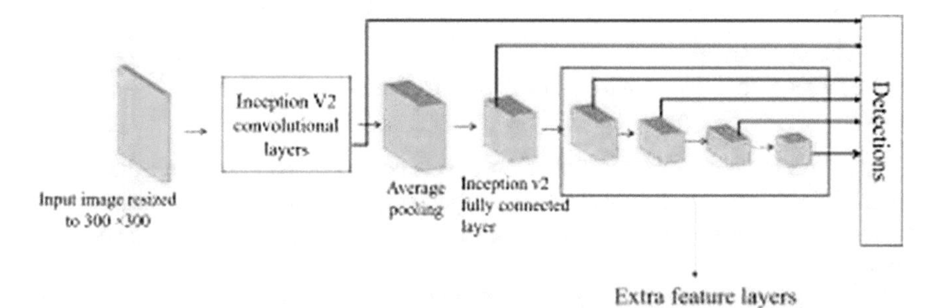

Fig. 17 Inception v2 architecture

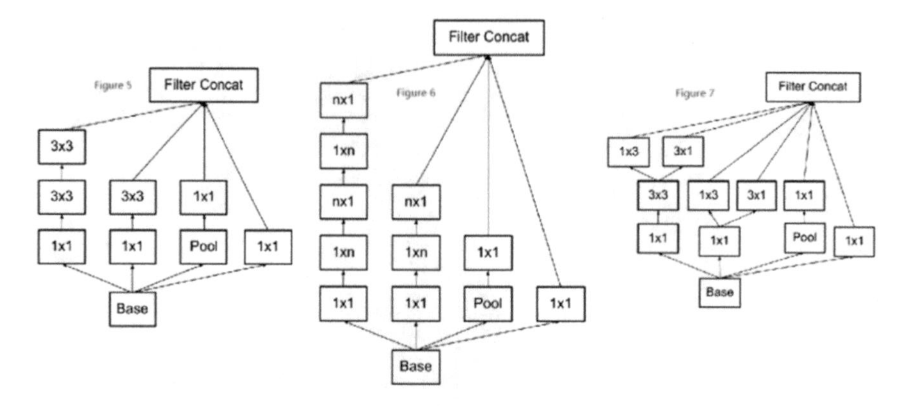

Fig. 18 Inception modules necessary for inception v2 (*Source* https://towardsdatascience.com/a-simple-guide-to-the-versions-of-the-inception-network-7fc52b863202)

size of 7 * 7 in the model. Spatial factorization has also been shown to boost the model's performance when used in asymmetric convolution. It was demonstrated here that n * n convolution may be substituted with n * 1 followed by 1 * n convolution, significantly lowering the computing cost of the model. Because of these features, this version outperformed the original Inception and v2. Specifically 3 × 3 is replacing 5 × 5 mainly to amplify space and increase performance (Fig. 18).

Shown above is the various types of factorization modules necessary for the V2 architecture. Where each module deals with bottleneck, parallelism, concatenation and increased computational speed.

4.9 DenseNet169

The next step in improving the depth of deep convolutional networks is to use dense nets, or densely connected convolutional networks. Dense Nets simplify the layer-to-layer communication pattern presented in prior architectures:

- Highway Systems.
- Residual Networks are a type of network that exists indefinitely.
- Networks of Fractals.

Contrary to popular assumption, Dense Nets require fewer parameters than traditional CNNs since they do not require learning redundant feature maps. Furthermore, many layers contribute relatively little and can be eliminated in numerous ResNet101 versions.

Because each layer has its own weights to learn, ResNet101 has a vast number of parameters. Dense Nets layers, on the other hand, are relatively narrow (for example, 12 filters) and only add a few additional feature-maps. Due to the previously explained information flow and gradients, another challenge with very deep networks was the difficulty in training them.

Dense Nets solve this problem by giving each layer direct access to the gradients from the loss function and the original input image (Fig. 19).

After applying a composite of operations, traditional feed-forward neural networks connect the output of the layer to the next layer. The equation for this would be:

$$x_l = H_l(x_{l-1})$$

This model immediately distinguishes itself from ResNet101. Dense Nets concatenate rather than sum the layer's output feature maps with the incoming feature maps. As a result, the equation reshapes into (Fig. 20):

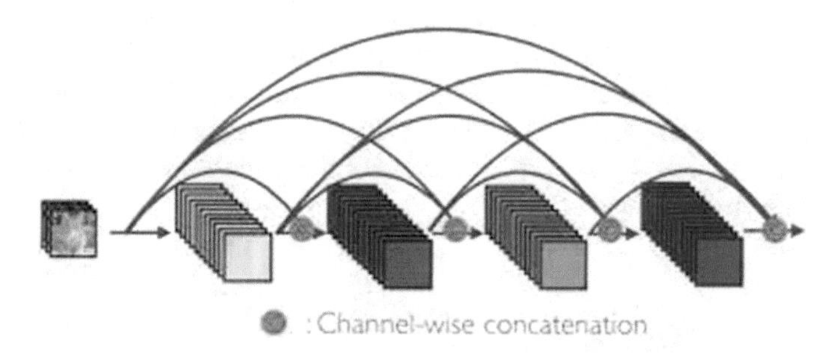

Fig. 19 Representation of Dense Net with 5 layers with an expansion of 4 (*Source* https://towardsdatascience.com/a-simple-guide-to-the-versions-of-the-inception-network-7fc52b863202)

Layers	Output Size	DenseNet-121	DenseNet-169	DenseNet-201	DenseNet-264
Convolution	112 × 112	7 × 7 conv, stride 2			
Pooling	56 × 56	3 × 3 max pool, stride 2			
Dense Block (1)	56 × 56	[1 × 1 conv, 3 × 3 conv] × 6	[1 × 1 conv, 3 × 3 conv] × 6	[1 × 1 conv, 3 × 3 conv] × 6	[1 × 1 conv, 3 × 3 conv] × 6
Transition Layer (1)	56 × 56	1 × 1 conv			
	28 × 28	2 × 2 average pool, stride 2			
Dense Block (2)	28 × 28	[1 × 1 conv, 3 × 3 conv] × 12	[1 × 1 conv, 3 × 3 conv] × 12	[1 × 1 conv, 3 × 3 conv] × 12	[1 × 1 conv, 3 × 3 conv] × 12
Transition Layer (2)	28 × 28	1 × 1 conv			
	14 × 14	2 × 2 average pool, stride 2			
Dense Block (3)	14 × 14	[1 × 1 conv, 3 × 3 conv] × 24	[1 × 1 conv, 3 × 3 conv] × 32	[1 × 1 conv, 3 × 3 conv] × 48	[1 × 1 conv, 3 × 3 conv] × 64
Transition Layer (3)	14 × 14	1 × 1 conv			
	7 × 7	2 × 2 average pool, stride 2			
Dense Block (4)	7 × 7	[1 × 1 conv, 3 × 3 conv] × 16	[1 × 1 conv, 3 × 3 conv] × 32	[1 × 1 conv, 3 × 3 conv] × 32	[1 × 1 conv, 3 × 3 conv] × 48
Classification Layer	1 × 1	7 × 7 global average pool			
		1000D fully-connected, softmax			

Fig. 20 Dense Net architectures (*Source* https://towardsdatascience.com/understanding-and-visualizing-densenets-7f688092391a)

$$x_l = H_l([x_0, x_1, \ldots, x_{l-1}])$$

Each architecture is composed from four Dense Blocks, each with a distinctive number of layers. DenseNet-121, for example, has [6, 12, 16, 24] layers in the four dense blocks, while DenseNet-169 has [6, 12, 32, 32] layers.

A 7 × 7 stride 2 Conv Layer is the first layer in the Dense Net architecture, followed by a 3 × 3 stride-2 MaxPooling layer. Following the fourth dense block is a Classification Layer, which accepts feature maps from all network tiers in order to classify them.

DenseNet-169 was chosen because, despite having 169 layers, it has a low parameter count compared to other models, and the architecture is well-suited to dealing with the vanish gradient problem. The last fully connected layer of the DenseNet-169 design was removed, and in its place, a 256 node fully connected layer, a 128 node FLC, and finally a 10 class FLC with softmax activation for the output were built.

ReLU activation was employed in everything but the final layer. FCLs were also subjected to batch normalisation and a 40% dropout.

Within each architecture, the Bottle Neck layers are also the convolution processes. This means that the 1 × 1 conv lowers the number of channels in the input, whereas the 3 × 3 conv convs the altered version of the input with the reduced number of channels instead of the input.

5 Results and Discussions

From the models we have developed, we have taken the best model from 50 epochs we have run and tested with Real-Time data we have obtained from CT scan centres. We have taken 300 images from 20 patient's CT chest scan images with an axial view of the lungs. This axial view is chosen with the radiographer's suggestion to observe systole and diastole of lungs with better view of infection spread.

After pre-processing the test data by applying filters and segmentation techniques, this data was tested using the 5 models we have developed i.e., Vgg16, ResnNet101, Mobile net, InceptionV2, DenseNet169. As mentioned above, the best model will be taken to test this data and results will be analysed using confusion matrix which gives us the True positives (TP), negatives (TN) and False positives (FP) and negatives (FN), which further helps in calculation result parameters like specificity, sensitivity and ROC curve. ROC (Receiver operator characteristic) curve is one of the evaluation metrics for binary classification problems. The area under this curve tells us how better the algorithm is in distinguishing the classes. Higher the area under the curve, better is the algorithm.

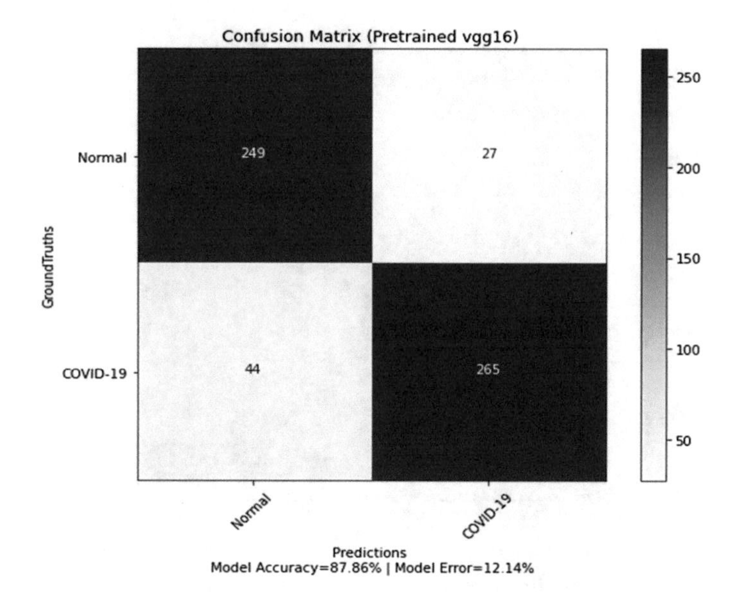

Fig. 21 Confusion matrix of VGG16

5.1 VGG16

We can observe that the model accuracy is 87.86% and the area under the curve for both the classes is 0.94. The ground truth visualization (given below) shows the infected area which is not precise (Heat map) and the ground truth, which is the precise infected area (Figs. 21 and 22).

5.2 Mobile Net

The model accuracy for Mobile Net is 75.56% which is which is less than VGG16, Inception net, ResNet and also DenseNet169. The area under the curve for mobile net is 0.84 which implies that it is not better at distinguishing the classes than VGG16 (Figs. 23, 24 and 25).

5.3 ResNet101

Among the models we have developed, ResNet101 showed the least accuracy of 90.82% and area under the curve of 0.97. We can also observe that the ground truth prediction from ResNet101 is highest compared to other models also it is best

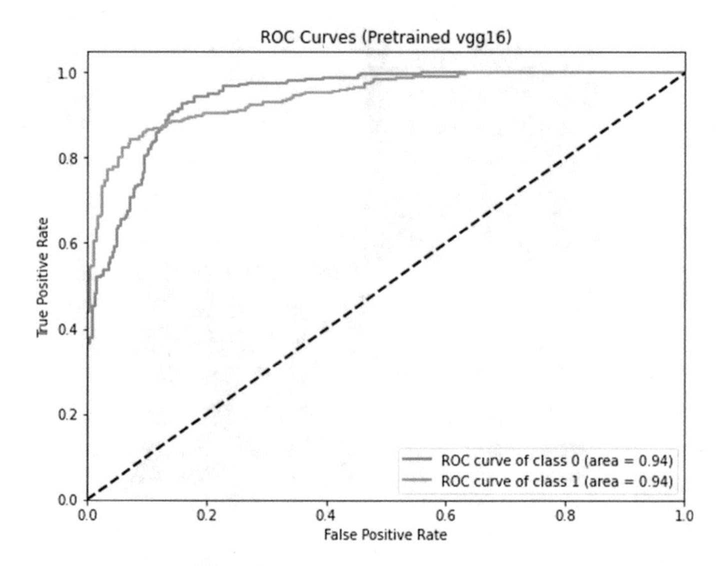

Fig. 22 ROC curve of VGG16

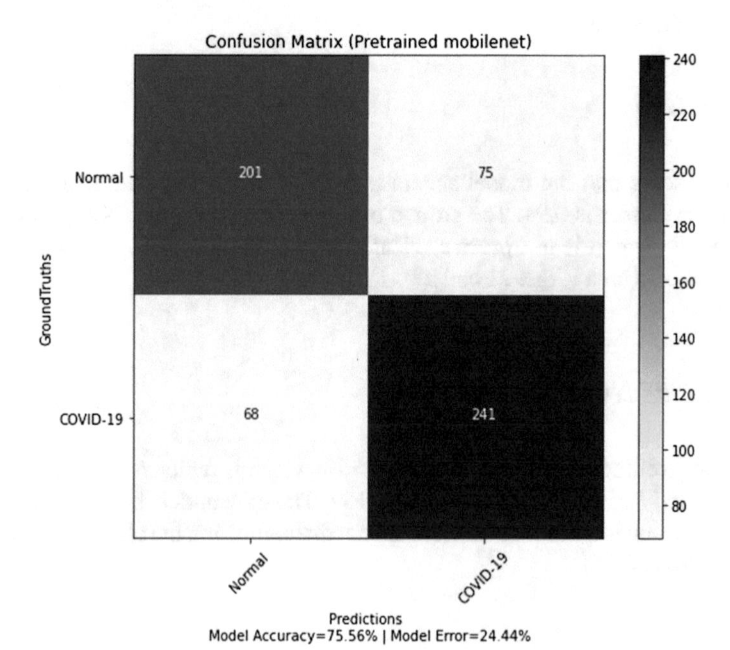

Fig. 23 Confusion matrix of mobile net

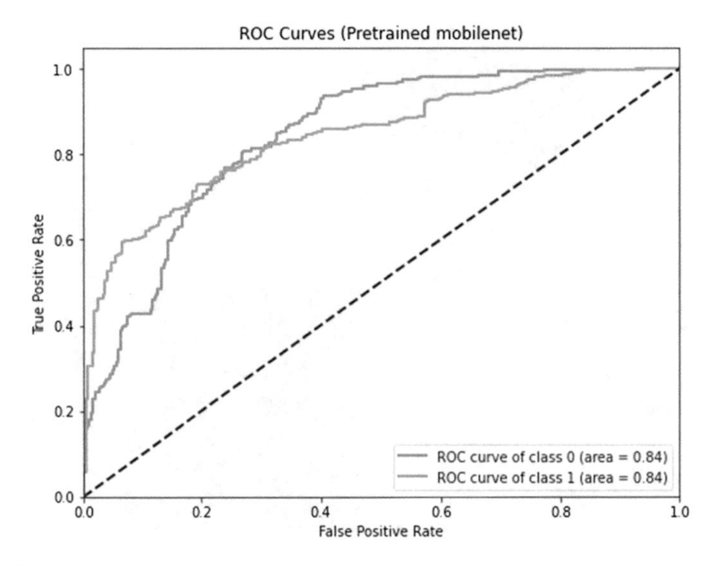

Fig. 24 ROC curve of mobile net

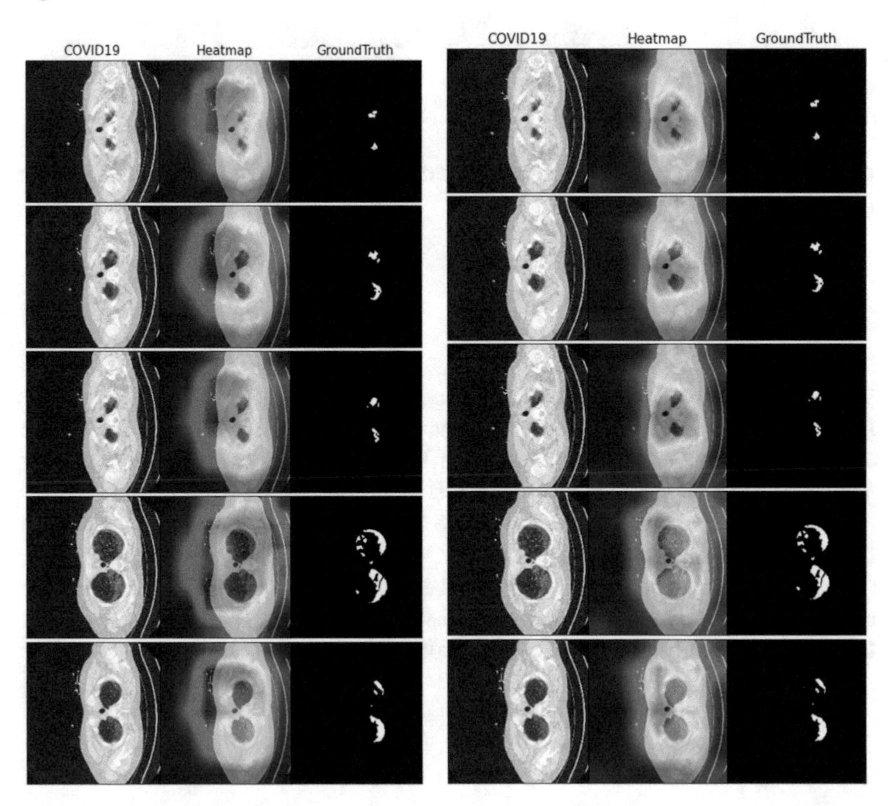

Fig. 25 Ground truth visualization of mobile net and VGG16

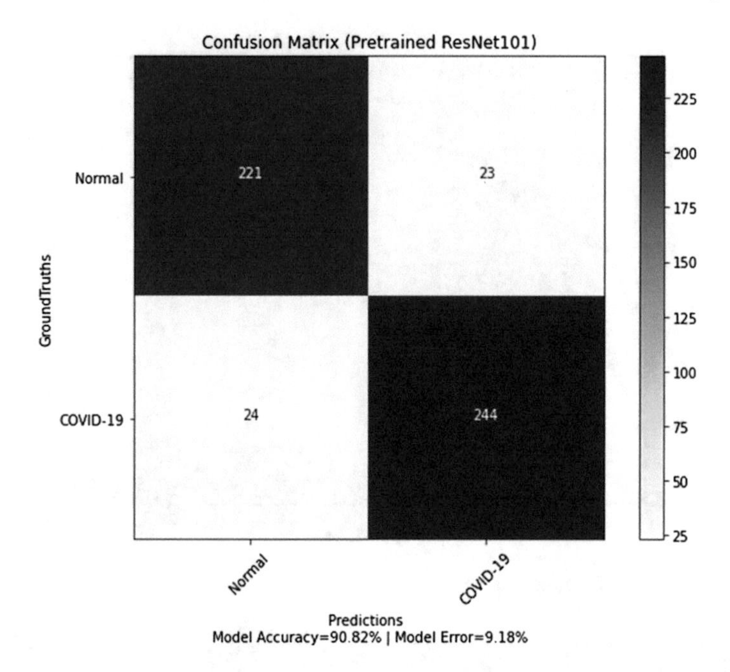

Fig. 26 Confusion matrix of ResNet101

model among those we've applied in terms of accuracy and displaying ground truths (Figs. 26 and 27).

5.4 Inception V2

We can observe that though the computational cost is less for inception V2 with the factorized convolution and dimension reduction, the model accuracy is 88.2%. The area under the curve is 0.95 for Inception V2. The ground truths and heat map of ResNet101 50 and inception v2 are given below (Figs. 28, 29 and 30).

The ground truth visualizations of these can be obtained by the mask applied on images as discussed in previous sections. We can observe that heat map for each model is different. The model segments the image data from the weights obtained after training. As mentioned earlier the best model weights will be considered for testing after running 50 epochs for each model.

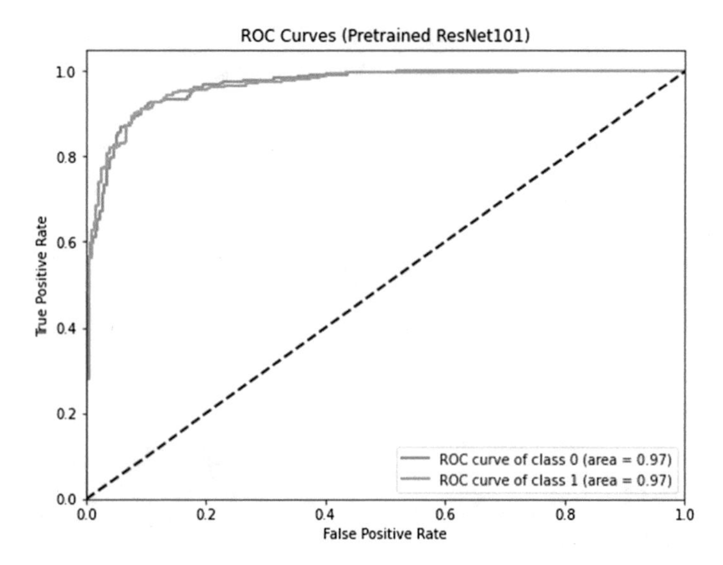

Fig. 27 ROC curve of ResNet101

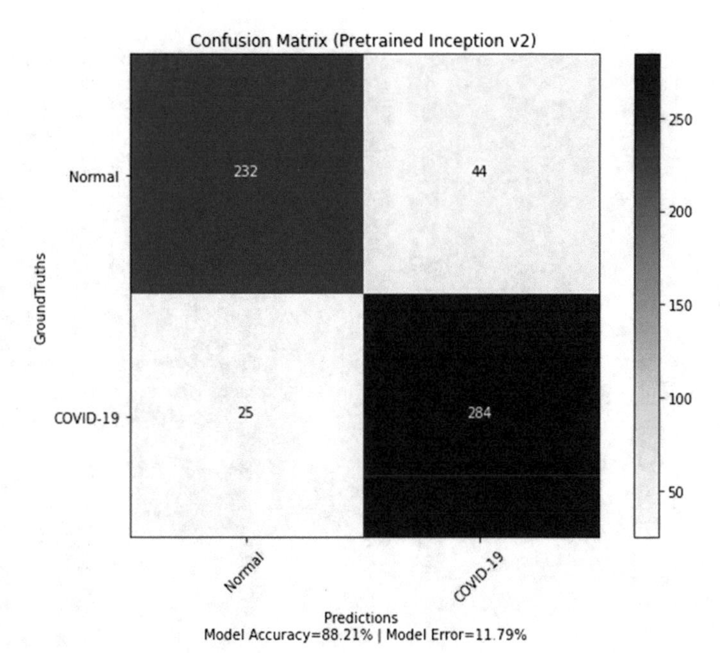

Fig. 28 Confusion matrix of Inception V2

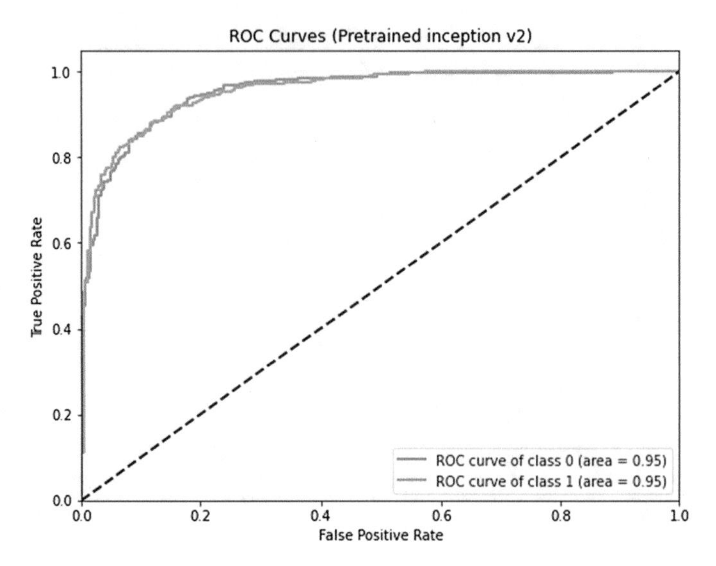

Fig. 29 ROC curve of Inception V2

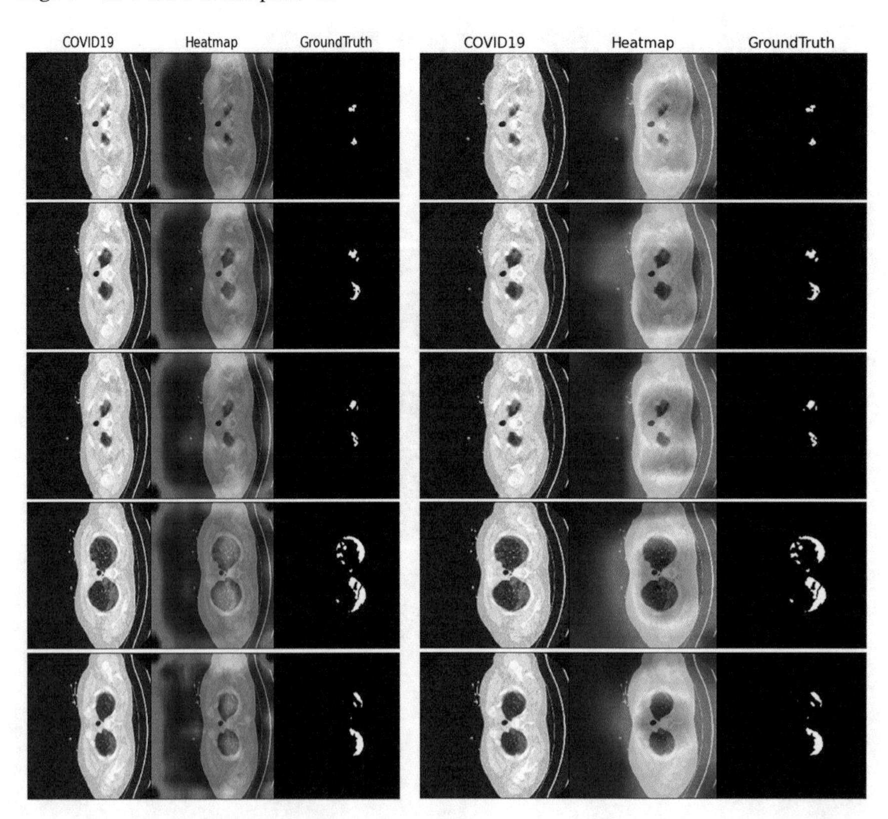

Fig. 30 Ground truth visualization of inception v2 and ResNet101101

5.5 Dense Net 169

With an accuracy of 89.57% and an area under the curve of 0.96, DenseNet169 has shown to be the second best model among those we've applied. We can also observe that the ground truth visualization of Dense net 121 is better than any other model with the precise detection of the infected part of the lungs. Specificity and sensitivity, which tell us the true positive and true negative rates of the models, are two further evaluation metrics employed in the examination of the algorithms in detecting Covid-19 from CT scans (Figs. 31, 32 and 33).

Formulae used for evaluation include:

$$\boldsymbol{Accuracy} = \frac{\left(\frac{TP}{TN}\right)}{(TP + FP + TN + FN)}$$

$$\boldsymbol{Precision} = \frac{\frac{TP}{TN}}{\left(\frac{TP}{TN}\right) + \left(\frac{FP}{FN}\right)}$$

$$\boldsymbol{Recall} = \frac{\frac{TP}{TN}}{\left(\frac{TP}{TN}\right) + \left(\frac{FN}{FP}\right)}$$

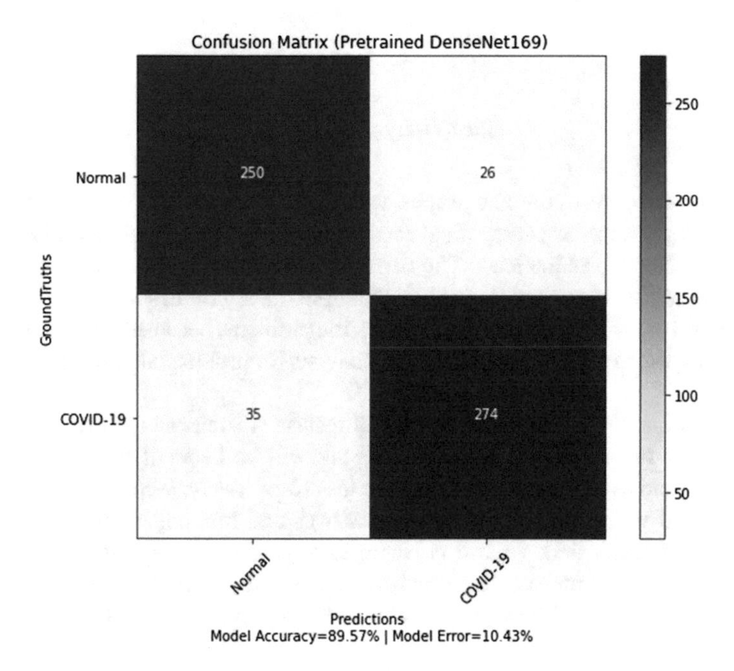

Fig. 31 Confusion matrix of Densenet169

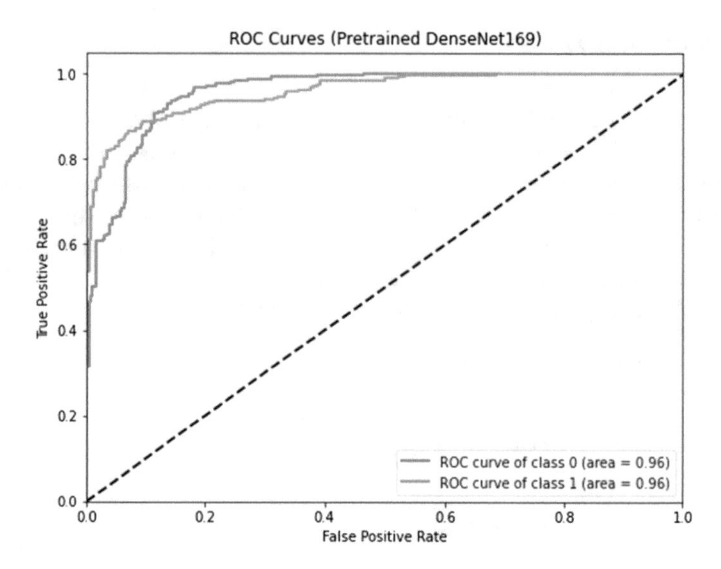

Fig. 32 ROC curve of Densenet169

$$f1_{score} = 2 * \frac{(Precision * Recall)}{Precision + Recall}$$

$$Sensitivity = \frac{TP}{TP + FN}$$

$$Specificity = \frac{TN}{TN + FP}$$

With the use of formulas and graphs, we can say from analysis of all the 5 models Res net 101 gives the best categorical accuracy and AUC-ROC among all the models and Mobile Net shows the least. The performance of the algorithm also depends on the dataset and it's pre-processing. For example, Inception has a newer version v3 but v2 is still considered best in object detection problems because of its architecture. That is why we have to perform different analysis for problem at hand (Figs. 34 and 35).

For each class in the dataset, evaluation measures such as f1 score, Recall, and Precision are quantified. Precision and Recall will be better if the false positives and false negatives predicted by model are less respectively. We can say that higher precision and recall, prediction of true positives and true negatives will be more. In order to quantify that, we use f1_score as a metric to do the analysis. It is the weighted average of metrics like Precision and Recall, where this micro averaging gives us better results with varying number of images in each class (Figs. 36 and 37).

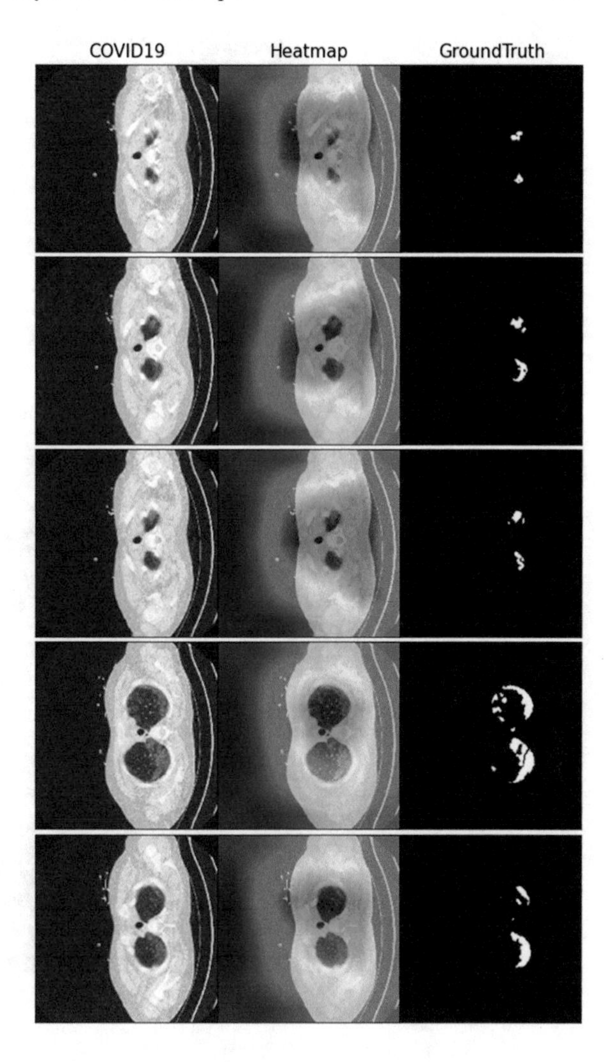

Fig. 33 Ground truth visualization of Densenet169

Sensitivity i.e.; True positive rate and Specificity which is also known as True negative rate are used for calculating the proportion of positive classes that are classified correctly and same with negative classes respectively. We can say that these are similar to Recall of that particular class but focuses more on the true predictions.

5.6 GUI

This interactive system of visual components is a windows application which conveys information and represents action that can be taken by the user. The same is used here to represent various AI models to detect COVID 19 and display whether or not the

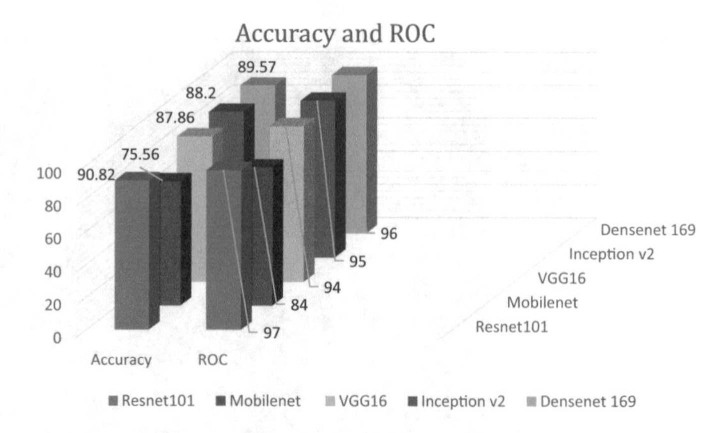

Fig. 34 Comparison of accuracy and ROC

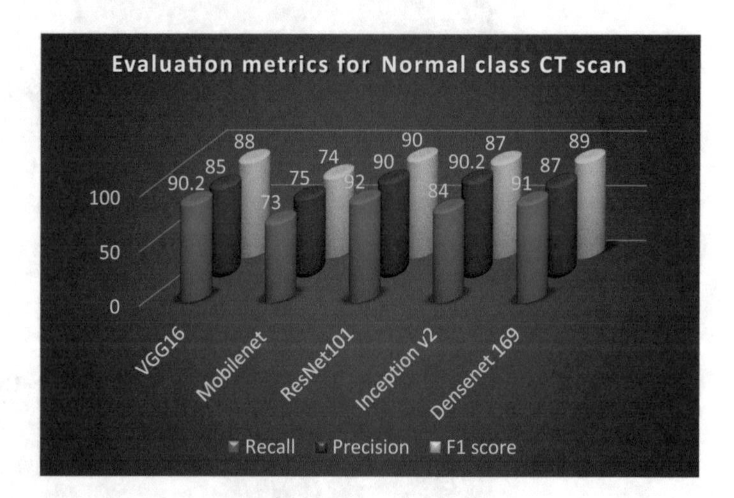

Fig. 35 Evaluation metrics for normal CT scan

scan contains a normal or an infected lung. Using this application, the user can also get a basic idea on how bad the infected lung is and to take precautions accordingly. The application looks like the Fig. 38.

This is how the application works: Firstly, the user should take an image from the device by using the browse option. The instructions to be followed are displayed on the left side of the application. The user will choose the image that can be either a normal lung CT scan and an infected lung CT scan. This is followed by the selection of models so as to detect the infected region (Fig. 39).

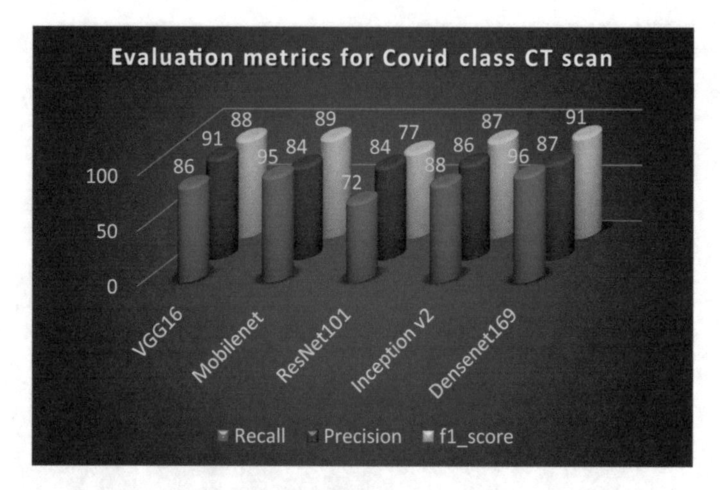

Fig. 36 Evaluation metrics of Covid CT scan

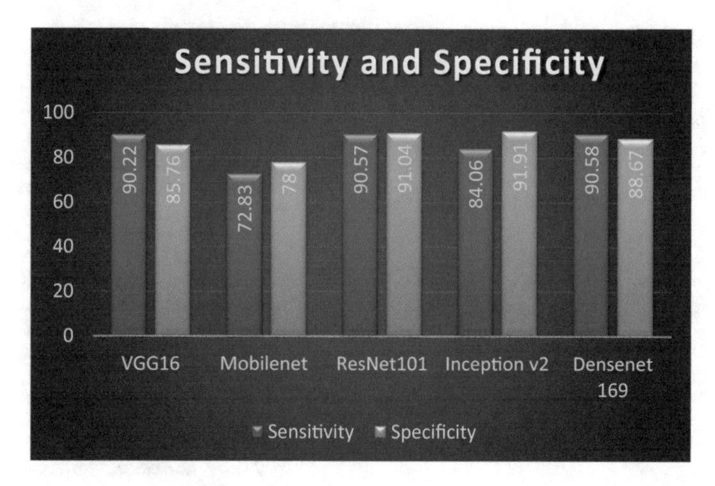

Fig. 37 Sensitivity and specificity

So even if the user does not select the model, Res Net 101 model will be selected by default because it gives a better accuracy of how much the lung is infected. The least accuracy is given by the model Mobile Net. Once the model is selected the user must click on detect COVID button. If there is presence of the disease then the output will indicate COVID positive, so the user should click on View Infected Region button to see how much infection is present (Fig. 40).

If there is absence of the disease then it will indicate COVID negative then the user can just click on exit button to close the app. The app also gives an option for COVID helpline which is indicated by the COVID button for further queries regarding the symptoms remedies to follow and vaccination related to the disease. The related

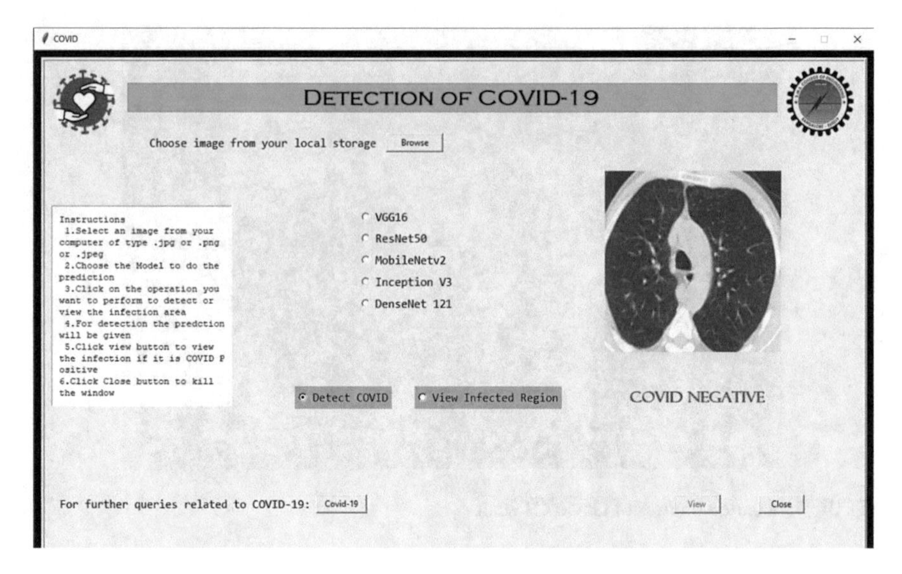

Fig. 38 Application overview

Fig. 39 Instructions box

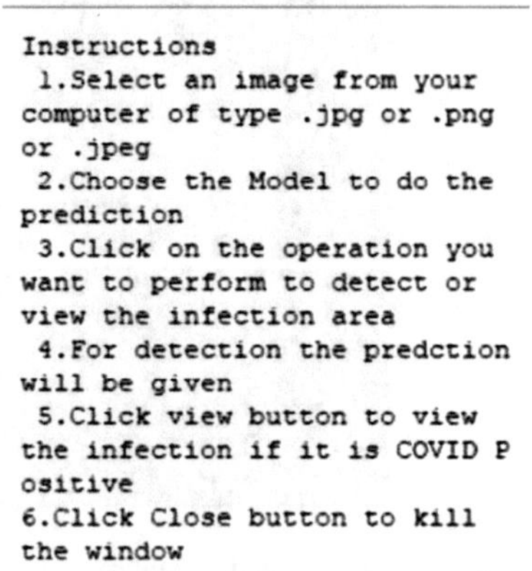

```
Instructions
 1.Select an image from your
computer of type .jpg or .png
or .jpeg
 2.Choose the Model to do the
prediction
 3.Click on the operation you
want to perform to detect or
view the infection area
 4.For detection the predction
will be given
 5.Click view button to view
the infection if it is COVID P
ositive
6.Click Close button to kill
the window
```

queries are from the WHO website which only gives relevant details to the user. This tool is in high demand in this pandemic for preliminary COVID detection (Figs. 41 and 42).

6 Conclusion

We used a variety of approaches to analyse covid detections from CT images in this work. It's difficult to train models with datasets gathered from AI scientists' segmentation databases and then test them with real-time data. The main goal of this

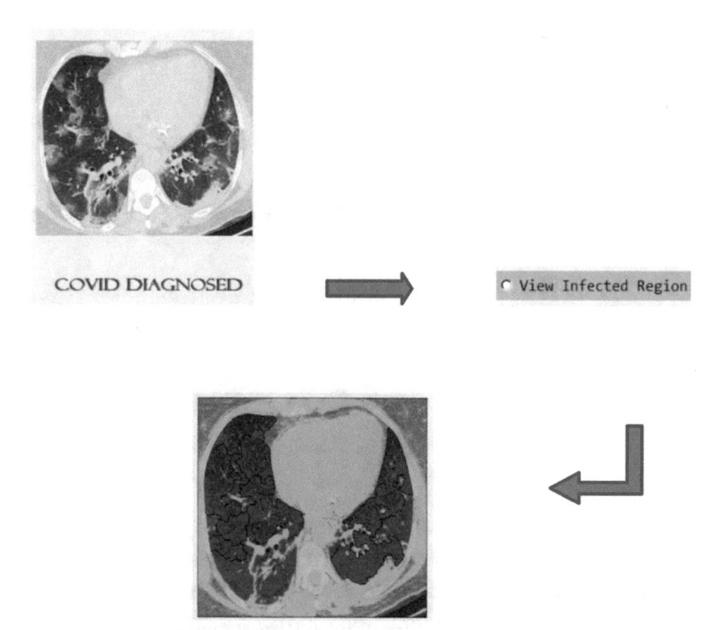

Fig. 40 View infected region

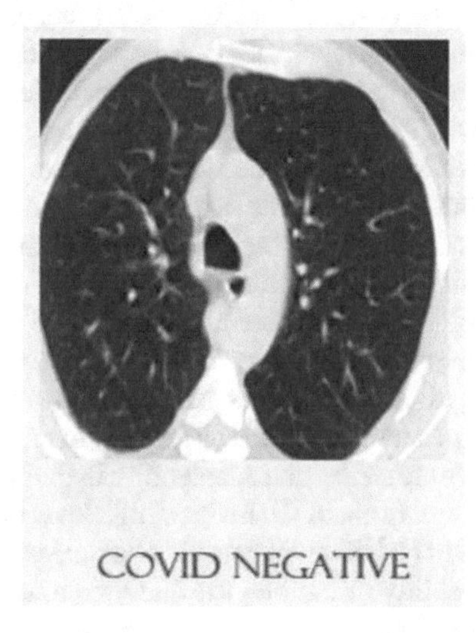

Fig. 41 COVID negative prediction

research is to find an algorithm that can predict the class of a raw CT scan without any processing, such that when a user interface (whether a web page or android app) is created for the general public, there is no need to pre-process data in order to obtain predictions.

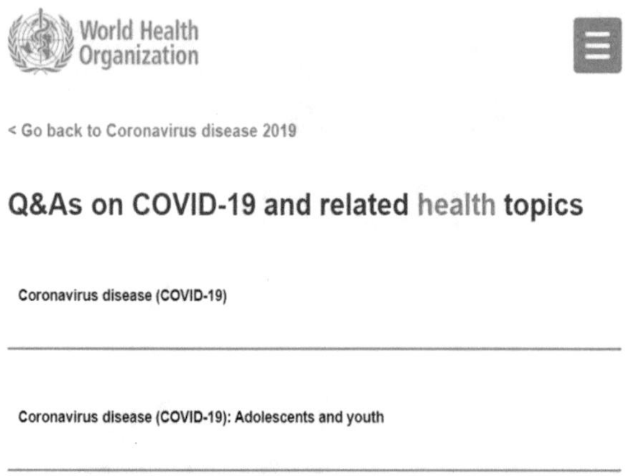

Fig. 42 WHO website (*Source* https://www.who.int/)

The 5 distinct deep learning algorithms were evaluated using a variety of parameters and metrics. Each algorithm's confusion matrix assists in the calculation of parameters. Because the dataset might be of variable sizes, the idea of ground truth versus prediction provides us the confusion matrix, which supports us in analysing the predictions for model behaviour particular to the class, where merely accuracy gives us an incorrect impression about it.

From the analysis we have observed that Res net 101 model gives the best accuracy with raw data and mobile Net gives the least. As we have mentioned earlier, model can be specific to that data used for training the model. Hence Res net 101 can be used to classify the raw CT scans. The patients shall be benefited from the user-friendly windows application because the CT scan report only offers the COVID grade on a scale of 0–6 but not the definite severity for grades 2–4. In such cases, this application will provide an exact result, as the project's primary objective is to assist the public/community with COVID-19 diagnosis utilising Artificial Intelligence and Computer vision technologies. As the technology is rapidly growing, it should be put to a good cause to help the community.

Healthcare and automobiles are two industries that have been severely impacted by the disruptions. This pandemic requires businesses to rethink their global supply chains in order to improve efficiency and adaptability in the case of an interruption in products availability. A supply chain GUI with a high level of flexibility can swiftly and efficiently manage impulsive or unexpected challenges or opportunities. These factors include the availability of radiologists and the demand for a tool for Covid preliminary self-diagnosis. Today's marketplaces are becoming more unpredictable, and a versatile supply chain can help the firm withstand potentially disastrous market changes in the healthcare industry.

References

1. Shi, F., Wang, J., Shi, J., Wu, Z., Wang, Q., Tang, Z., He, K., Shi, Y., Shen, D.: Review of artificial intelligence techniques in imaging data acquisition, segmentation and diagnosis for COVID-19. IEEE Rev. Biomed. Eng. (2020). https://doi.org/10.1109/RBME.2020.2987975
2. Chung, M., Bernheim, A., Mei, X., et al.: CT imaging features of 2019 novel coronavirus (2019-nCoV). Radiology **295**(1), 202–207 (2020)
3. Ishida, S.: Perspectives on supply chain management in a pandemic and the post-COVID-19 era. IEEE Eng. Manag. Rev. **48**(3), 146–152 (2020). https://doi.org/10.1109/EMR.2020.3016350
4. Maiti, A., Shilpa, R.G.: Developing a framework to digitize supply chain between supplier and manufacturer. In: 2020 5th International Conference on Computing, Communication and Security (ICCCS) (2020), pp. 1–6. https://doi.org/10.1109/ICCCS49678.2020.9277211
5. Wu, Y.-H., et al.: Jcs: an explainable covid-19 diagnosis system by joint classification and segmentation (2020). arXiv:2004.07054
6. Wang, D., Hu, B., Hu, C., et al.: Clinical characteristics of 138 hospitalized patients with 2019 novel coronavirus-infected pneumonia in Wuhan, China. JAMA (2020) [Epub ahead of print]
7. Ai, T., Yang, Z., Hou, H., et al.: Correlation of chest CT and RT-PCR testing in coronavirus disease 2019 (COVID-19) in China: a report of 1014 cases. Radiology **26**, 200642 (2020) [Epub ahead of print]
8. Fang, Y., Zhang, H., Xie, J., et al.: Sensitivity of chest CT for COVID-19: comparison to RT-PCR. Radiology **19**, 200432 (2020) [Epub ahead of print]
9. Chapman, W.W., Bridewell, W., Hanbury, P., Cooper, G.F., Buchanan, B.G.: A simple algorithm for identifying negated findings and diseases in discharge summaries. J. Biomed. Inform. **34**(5), 301–310 (2001)
10. Sølund, T., Buch, A.G., Krüger, N., Aanæs, H.: A large-scale 3D object recognition dataset. In: 2016 Fourth International Conference on 3D Vision (3DV), pp. 73–82 (2016)
11. Gururaj, C., Tunga, S.: AI based feature extraction through content based image retrieval. J. Comput. Theor. Nanosci. **17**(9–10), 4097–4101 (2020). ISSN: 1546-1955
12. Chen, N., Zhou, M., Dong, X., et al.: Epidemiological and clinical characteristics of 99 cases of 2019 novel coronavirus pneumonia in Wuhan, China: a descriptive study. Lancet **395**(10223), 507–513 (2020)
13. Holshue, M.L., DeBolt, C., Lindquist, S., et al.: First case of 2019 novel coronavirus in the United States. N. Engl. J. Med. **382**(10), 929–936 (2020)
14. Zhang, J., Xie, Y., Li, Y., Shen, C., Xia, Y.: COVID-19 screening on chest X-ray images using deep learning based anomaly detection (2020). arXiv:2003.12338. [Online]. http://arxiv.org/abs/2003.12338
15. Li, Q., Guan, X., Wu, P., et al.: Early transmission dynamics in Wuhan, China, of novel coronavirus-infected pneumonia. N. Engl. J. Med. (2020) [Epub ahead of print]
16. Yan, K., Wang, X., Lu, L., Summers, R.M.: DeepLesion: automated mining of large-scale lesion annotations and universal lesion detection with deep learning. J. Med. Imag. **5**, Art. no. 036501 (2018)
17. Mei, X., et al.: Artificial intelligence-enabled rapid diagnosis of patients with COVID-19. Nat. Med. **26**(8), 1224–1228 (2020)
18. Shi, F., et al.: Review of artificial intelligence techniques in imaging data acquisition, segmentation and diagnosis for COVID-19. IEEE Rev. Biomed. Eng. https://doi.org/10.1109/RBME.2020.2987975
19. Narin, A., Kaya, C., Pamuk, Z.: Automatic detection of coronavirus disease (COVID-19) using X-ray images and deep convolutional neural networks (2020). arXiv:2003.10849
20. Gururaj, C.: Proficient algorithm for features mining in fundus images through content based image retrieval. In: IEEE International Conference on Intelligent and Innovative Computing Applications (ICONIC-2018), 6–7 Dec 2018, pp. 108–113, Plaine Magnien, Mauritius. https://doi.org/10.1109/ICONIC.2018.8601259. ISBN 978-1-5386-6476-6

21. Greenspan, H., van Ginneken, B., Summers, R.M.: Guest editorial deep learning in medical imaging: overview and future promise of an exciting new technique. IEEE Trans. Med. Imaging **35**(5), 1153–1159 (2016). https://doi.org/10.1109/TMI.2016.2553401
22. Tabik, S., et al.: COVIDGR dataset and COVID-SDNet methodology for predicting COVID-19 based on chest X-ray images. IEEE J. Biomed. Health Inform. **24**(12), 3595–3605 (2020). https://doi.org/10.1109/JBHI.2020.3037127
23. Gururaj, C ., Jayadevappa, D., Tunga, S.: Content based image retrieval system implementation through neural network. IOSR J. VLSI Signal Process. (IOSR-JVSP) **6**(3), 42–47 (Ver. 3) (2016). https://doi.org/10.9790/4200-0603034247. e-ISSN: 2319-4200, p-ISSN No.: 2319-4197
24. Wang, X., Peng, Y., Lu, L., Lu, Z., Bagheri, M., Summers, R.M.: Chest X-ray 8: hospital-scale chest X-ray database and benchmarks on weakly-supervised classification and localization of common thorax diseases. In: Proceedings of the IEEE Conference on Computer Vision and Pattern Recognition, pp. 3462–3471 (2017)
25. Ghezelghieh, M.F., Kasturi, R., Sarkar, S.: Learning camera viewpoint using CNN to improve 3D body Pose estimation. In: 2016 Fourth International Conference on 3D Vision (3DV), pp. 685–693 (2016)
26. Irvin, J., et al.: CheXpert: a large chest radiograph dataset with uncertainty labels and expert comparison. In: Proceedings of the AAAI Conference on Artificial Intelligence, pp. 590–597 (2019)
27. Johnson, A.E.W., et al.: MIMIC-CXR-JPG, a large publicly available database of labeled chest radiographs (2019). arXiv:1901.07042
28. Xu, Z., Elomri, A., Kerbache, L., El Omri,A.: Impacts of COVID-19 on global supply chains: facts and perspectives. IEEE Eng. Manag. Rev. **48**(3), 153–166 (2020). https://doi.org/10.1109/EMR.2020.3018420.
29. https://towardsdatascience.com/understanding-and-visualizing-densenets-7f688092391a
30. https://neurohive.io/en/popular-networks/vgg16/

Integrated Supply Chain Process for Dairy Management

Rabie A. Ramadan, Ramya Govindaraj, A. Praveen, and Kumaresan Perumal

Abstract Managing the works manually for the dairy management has become tough task. A dairy management System helps in reducing the manual work such as a regular daily activity can be automated thus making the tasks easier. When the time is reduced, we can work more that helps in increasing the profit. Some of daily works such as milk collections from vendors, sales to the customer and plant and all the dairy-related processes can be easily managed with the help of Dairy Management System. Also, we can track the movements of the products like manufactured products, raw materials from the warehouse to the respected customers. All those process and activities can be managed with the help of ERP systems. Also, activities like accounting, forecasting, locating the warehouse, scheduling, reporting can be done by the management system. It also helps in Cattle management, Employee management, report analysis and forecasting, accounting and finance, supply chain and a lot of various ERP related activities.

Keywords Accounting and finance · Sales forecasting · Reporting · Supply chain · Locating warehouse

R. Govindaraj (✉) · A. Praveen · K. Perumal
School of Information Technology and Engineering, Vellore Institute of Technology, Vellore, Tamil Nadu, India
e-mail: ramya.g@vit.ac.in

A. Praveen
e-mail: praveen.a2018@vitstudent.ac.in

K. Perumal
e-mail: pkumaresan@vit.ac.in

R. A. Ramadan
Computer Engineering Department, Cairo University, Cairo, Egypt
e-mail: rabie@rabieramadan.org

K. Perumal et al. (eds.), *Innovative Supply Chain Management via Digitalization and Artificial Intelligence*, Studies in Systems, Decision and Control 424,
https://doi.org/10.1007/978-981-19-0240-6_10

1 Relevant Literature

Ravi et al. [1] has discussed in this paper about the importance of integrating ICT based ERP in the dairy management system. By integrating the ICT, it can improve the co-ordination between the milk cooperatives and cattle farmers. They have conducted the research in four districts of Gujarat and collected the primary data from the cattle farmers. With the reference to the collected data, they have identified the challenges and problems and also the requirements needed by the milk co-operatives to make their proposed model. The milk co-operatives needed the improvement in co-ordination between them Thus a centralized level of ERP provided with many details is needed. The information such as the number of cattle in each husbandry, Husbandry wise milk production report, each cattle farmer loan account detail, per cattle milk production, cattle wise medical expense detail. The milk co-operatives should also have a farmer relationship management module. It should centrally integrate and monitor all the cattle farmers and cattle farms. The farmers face the difficulties such as keep track of vaccination time schedule, having the number of lactation days and number of dry days detail, finding a new veterinary doctor. These problems are sorted out by the provided ICT based ERP he concluded.

Sengupta [2] has discussed about the importance of robotics in manufacturing the dairy product in the dairy industry. As a huge demand raises for milk quotes in Asian and African countries it has put the margins under pressure. Also, the customers and the government agencies are demanding more which become more complex to produce. Monitoring the regulations, documenting the customers information, integration with payment system will provide a strong backbone to the arising problems. By using ERP, dairy companies are choosing an integrated system that will allow them to achieve far more control of their processes [3, 4]. All data streams converge in one system, giving them better insight and a comprehensive overview, and allowing them to determine their position with a single click. In addition, the system will provide dairy companies a far more future-oriented and solid backbone. This is very useful, as they are receiving more and more information to process from the supply chain. Data from the production process, the supply chain, the contracts, taxes, and much, much more, will all be stored and processed centrally. In order to stay competitive and to maintain the margins, digital transformation of the dairy industry is a must. Integrated systems and harvesting various kinds of data from the supply chain will allow cost reduction, allow the complexity to be alleviated and allow businesses to be more responsive operating on a volatile market.

Patil et al. [3] has discussed the depth of ERP in the dairy cooperatives. With ERP we can achieve lot of things it is more profitable for the organization. These are tangible and intangible benefits like—Web Based—As recent ERP' s is web-based data is recorded of total enterprises irrespective of physical location hence total picture of organization is available One check introduction as any location with help of web-based ERP. Management Information System—As all data is recorded at source. We can generate daily reports for top management based on defined priority.

SMS can be sent to respective authority. Today's up to date comparison with last year—Today's milk collection and sale up to date from accounting year report can be generated. It also provides comparison with last year. Automaton—Most of plants operations can be atomized e.g., temperature recording or opening of valves at certain temp/ level etc. Alert and messages—while recording data on-line based on pre-defined rules alerts can be given or SMS can be generated e.g., if customer demand of SMP for 1 ton SMS to marketing supervisor, if it is for 5 tons SMS to marketing manager if is for 10 tons SMS Managing Director and SMS to the chairman for above 10 tones. Validation and Verification—As ERP considered as integrated data and data enter into system at source its validation is done while entering and its proper verification is done to avoid further effects of data [5, 6].

2 Short Description

2.1 ERP Tool

Dolibarr ERP & CRM is an open-source and free software package to manage companies, freelancers or foundations. We can say Dolibarr is an ERP or CRM (or both depending on activated modules). It's an Opensource project built by the addition of modules (you only enable the features you need), on a WAMP, MAMP or LAMP server (Apache, MySQL, PHP for all Operating Systems). Dolibarr was developed to try to offer an ERP and CRM suite with the main goal of simplicity:

- Simple to install
- Simple to use
- Simple to develop

2.2 Software Requirements

- Any one Browser (Microsoft Edge, Google Chrome, Mozilla Firefox)
- DOLIBARR 12.0.4 installed as localhost
- Apache Server 2.4.41 or higher
- Maria DB 10.4.11 or higher
- Python v3.6 or higher
- Google Collaboratory for working on python as IPYNB file
- KERAS and TENSORFLOW for CNN classification
- GEOPANDAS for analyzing Geospatial Data
- Linear Regression Model for Sales Forecasting
- Microsoft Visio (for Charts and Diagrams)

2.3 Hardware Requirements

- Laptop with any operating system pre-installed, minimum of 4 GB RAM
- Windows 7 or above Operating System.
- Internet Connectivity, with moderate speed.

2.4 Process Operation

The system is mainly focused on dairy management and their product management. When the integrated ERP system is built for Dairy Management System, we can store all the data and reports in the database and can be utilized by ERP system. The system is built to make the user interface easier and easily accessible to the data and reports needed.

3 System Architecture

List of Process operation of conventional dairy management system is shown in Fig. 1 and its components are the following,

- Buying the Raw Materials from the Vendors
- Storing them in the vacant warehouse
- Manufacturing the Products with the Raw Material available
- Making the Products available for Sale
- Allowing Customers to buy the products
- Making the shipment of the products from the warehouse
- Generating the invoice for the sales order
- Collecting the sales amount and transferring to the Bank Account
- Regularly check for the vacancy in the warehouse

When ERP is integrated to the Dairy Management system, we can automate all the process and the reports of those process can also be regularly maintained [7]. The process operation of the integrated dairy management system includes [8–12]:

1. Finding the Hotspot for locating the Warehouse so the distribution is made easier.
2. Maintaining the records of the items and their quantity in the warehouse so buying the raw materials can be easily intimated
3. Make available the list of Vendors from the previous purchases so that we can choose the best vendor and nearby vendor to the located warehouse.
4. Shows the previous purchase orders and their prices.
5. Maintaining the records of the Purchase Order of the dairy products and generating the invoice for the accounting purpose.

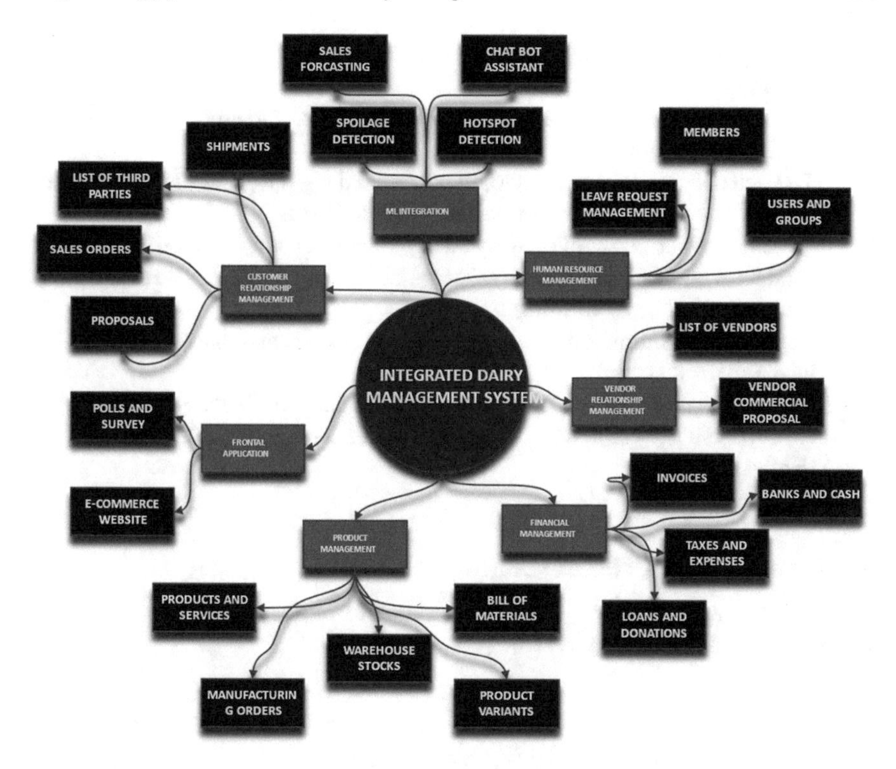

Fig. 1 System Design

6. Bank account can be created and linked so that the purchase orders can be paid within the ERP system itself
7. Navigating the Purchase Orders where to store the number of items in which warehouses.
8. Updating the number of items in the warehouse so that the stocks are regularly checked and no confusion prevails with Sales Order
9. Creating the Bill of Materials for the products such as Cheese and Butter so that those products can be manufactured [13].
10. For Manufacturing the orders, the raw materials are taking from the warehouse nearby and processed for Manufacturing
11. Whenever the items are taking, kept and transferred from the warehouse it will be automatically recorded.
12. After Manufacturing Order, the new products are transferred to the warehouse and list gets updated.
13. All the products that are available for sale should be estimated for the Selling Price with Inclusive or Exclusive of GST
14. Making the product available for sales and let the customers open for buying the products by E-Commerce Website.

15. Every purchase made by customers and 3rd parties are recorded and invoice is generated.
16. All the money obtained from the sales order can be directly transferred to the Bank account linked.
17. The items in the Warehouse should be checked regularly because the products are related to dairy and it can be easily spoiled [14].
18. Warehouse can be monitored by Image Classification of Good and Spoiled Product with CNN provided by Computer Vision (CV). Live webcam can be feed for real-time monitoring.
19. Reports of the Inflow and Outflow of the company is maintained thus sales can be maintained stably.
20. Even the sales can be forecasted with the help of Machine Learning by REGRESSION techniques. Thus, the orders and stocks can be maintained in advance.

4 Event Driven Process Chain

This process is shown in Fig. 2 [15].

5 Customer Relationship Management

5.1 Third Parties

Third Parties are Prospects, Customers, Vendors are the people who involves in relationship between the company by proposing the orders, selling the orders and buying the orders and raw materials shown in Fig. 3. Dolibarr shows a statistic of no. of Prospects, Customers and Vendors in the Donut graph. Also, the recently active 15 Third Parties are mentioned here.

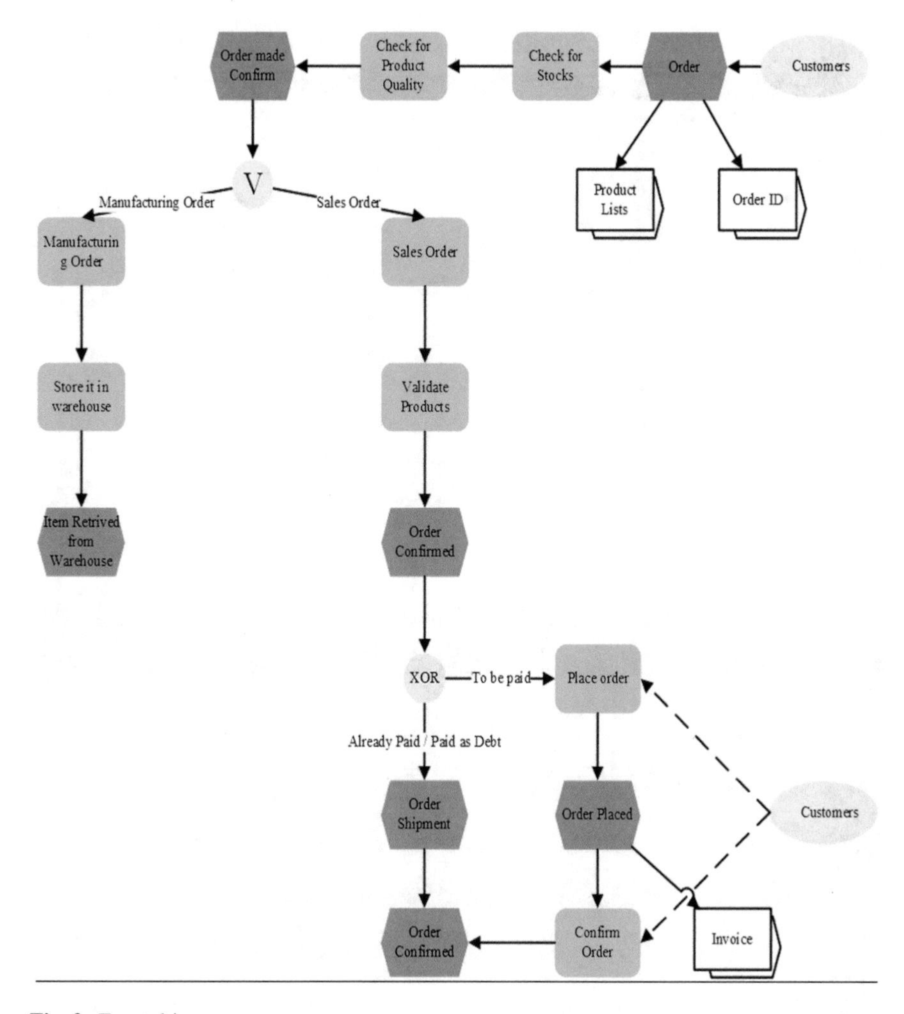

Fig. 2 Event driven process

5.2 List of Prospects/Customers

The list of customers and prospects are mentioned in Fig. 4. Customers are the person who buys the company' product and prospects are the one who is interested in buying the company's product. Here we can sort the list by the name, location and zip code. Also, the status of the Customers/Prospects are mentioned here.

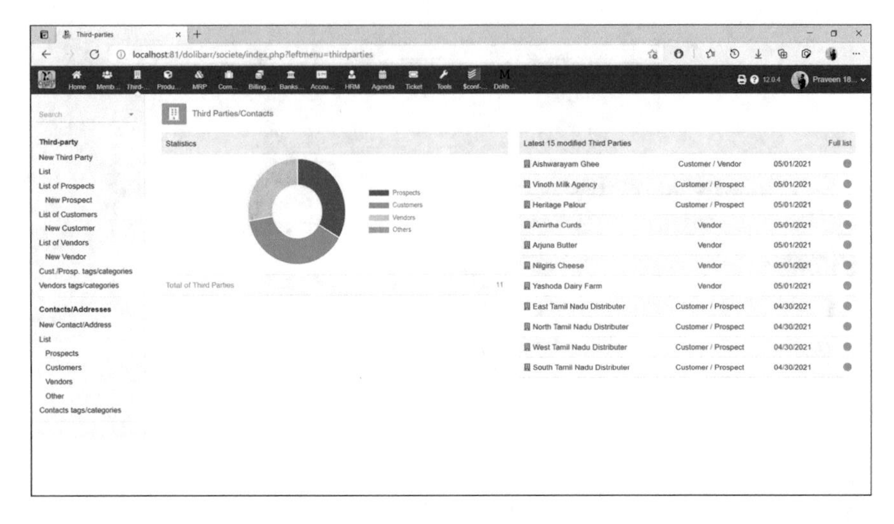

Fig. 3 Third parties

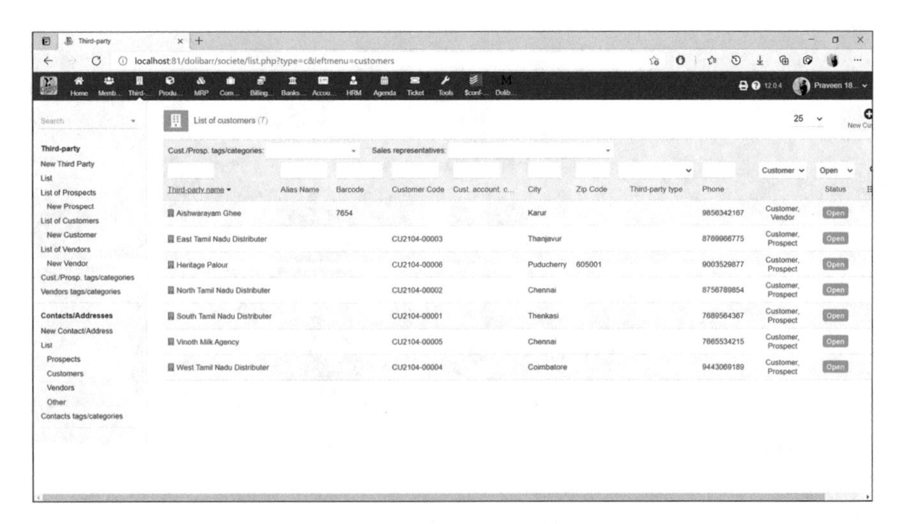

Fig. 4 List of customers

5.3 Prospects/Customers

The details of every prospect and customer are stored. It includes the name of the customers, address, contact number, customer code and it is shown in Fig. 5. Also, the details of the latest sales order, shipment, invoices are included with the status of those. The total amount of Proposals, Orders, Invoices and Current outstanding are also mentioned.

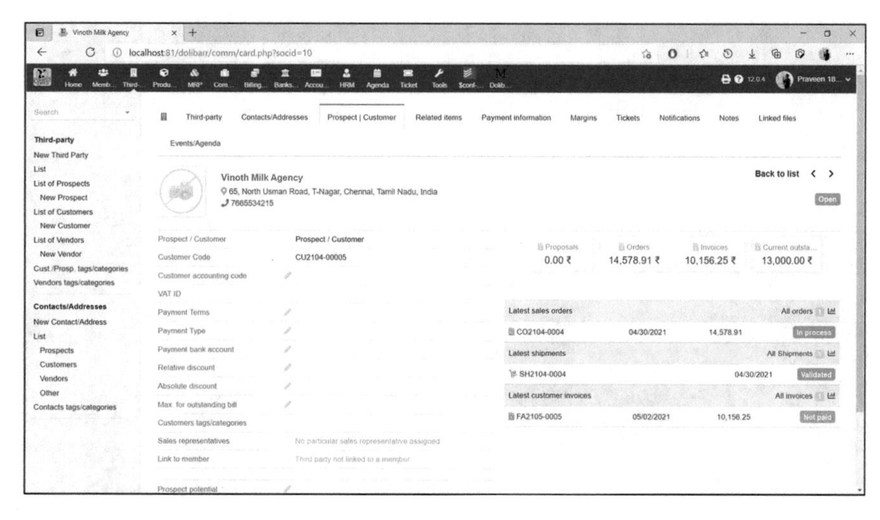

Fig. 5 Details of customers

5.4 List of Proposals

The list of proposal can also be viewed in Fig. 6. Here we can sort out the proposal with respect to the Third Parties, Start Date, End Date, Amount. Also, the status of the proposal whether it is pending, validated, approved. The Author who made this proposal is also mentioned here.

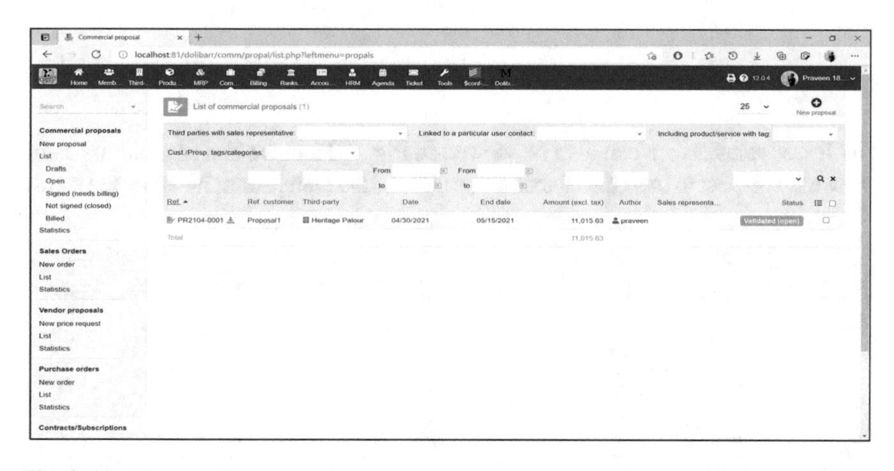

Fig. 6 List of proposals

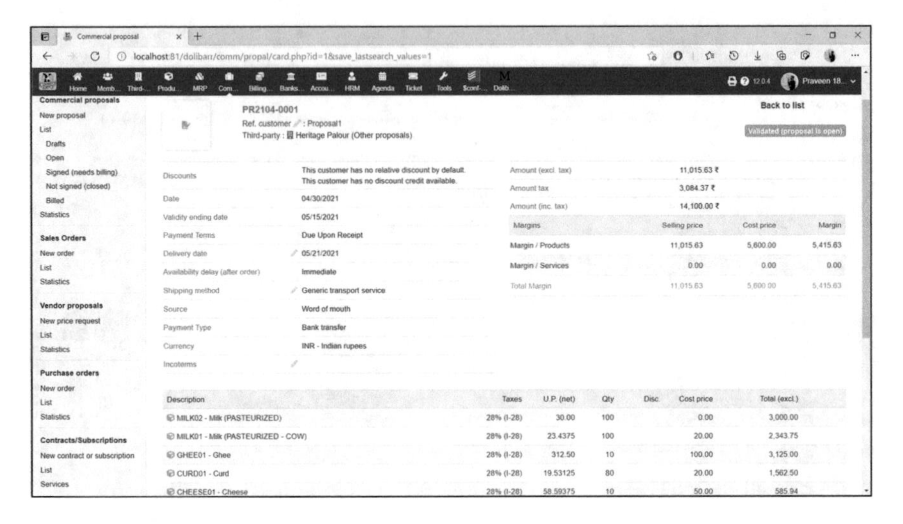

Fig. 7 Details of proposals

5.5 *Proposals*

The details of proposals are mentioned in Fig. 7. The details such as Proposal name, to whom the proposal made, date of the proposal, payment terms are all mentioned. The Product descriptions along with the quantity and price of those products along with the taxes are also mentioned.

5.6 *List of Sales Orders*

The list of sales orders can also be viewed in Fig. 8. Here we can sort out the sales order with respect to the Third Parties, Start Date, End Date, Amount. Also, the status

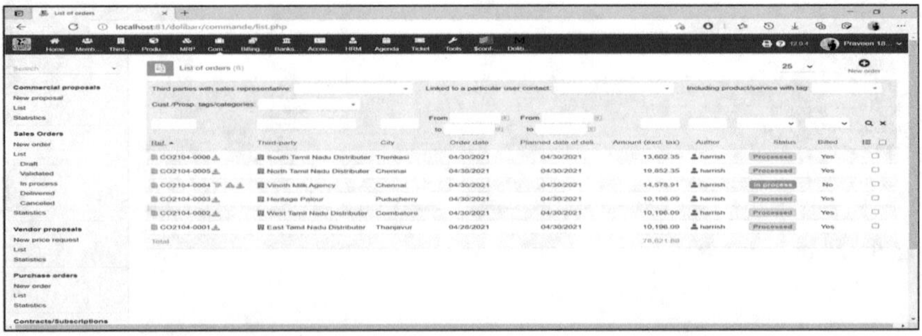

Fig. 8 List of sales order

of the sales orders whether it is not started, in process, Processed. The Author who made this sales order is also mentioned here.

5.7 *Sales Order*

The details of sales order are mentioned in Fig. 9. The details such as sales order id, the customer of the sales order, date of the ordered, payment terms are all mentioned. The Product descriptions along with the quantity and price of those products along with the taxes are also mentioned.

5.8 *List of Shipments*

The list of shipments can also be viewed. Here we can sort out the sales order with respect to the third parties, delivery date, address, location. Also, the status of the sales orders whether it is not validated, in process, Validated. The shipment here is mentioned along with the sales order id. The details of shipments are mentioned here. Every shipment is the attachment of the sales order.

Thus, all shipments are mentioned with the respective sales order. The details such as shipment id, delivery date of the ordered, shipping methods are all mentioned. The Product descriptions along with the quantity and the source warehouse where it is available are also mentioned.

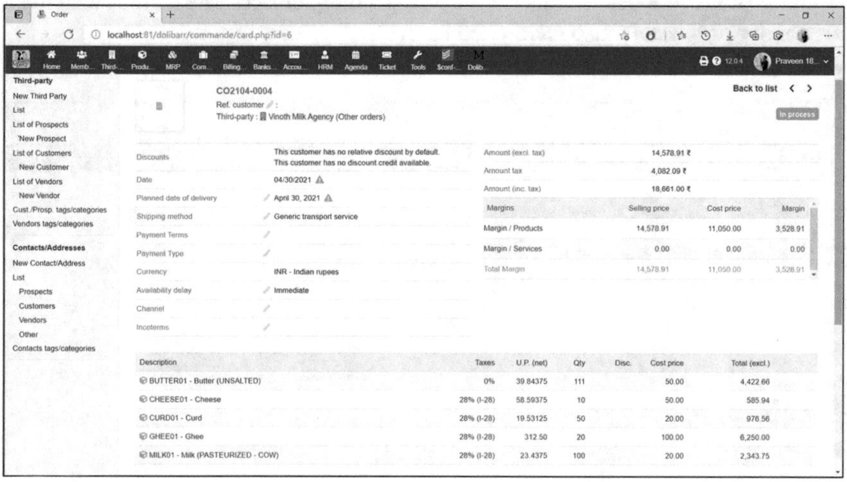

Fig. 9 Details of sales order

6 Vendor Relationship Management

6.1 *List of Vendors*

The list of vendors is mentioned here. Vendors are the person whom the company buys the products and raw materials. In simple the suppliers of the company are mentioned as vendors. Here we can sort the list by the name, location and zip code. Also, the status of the Vendors is mentioned here (Fig. 10).

6.2 *Vendors*

The details of every vendor are stored and it is shown in Fig. 11. It includes the name of the vendors, address, contact number, vendor code. Also, the details of the latest purchase order, Supplier proposals, vendor invoices are included with the status of those. The total amount of Price Request, Orders, Invoices and Current outstanding are also mentioned.

6.3 *List of Vendor Commercial Proposal*

The list of vendor proposal can also be viewed in Fig. 12. Here we can sort out the vendor proposal with respect to the vendor name, start date, end date, amount. Also, the status of the proposal whether it is open, pending, closed. The Author who made this vendor proposal is also mentioned here.

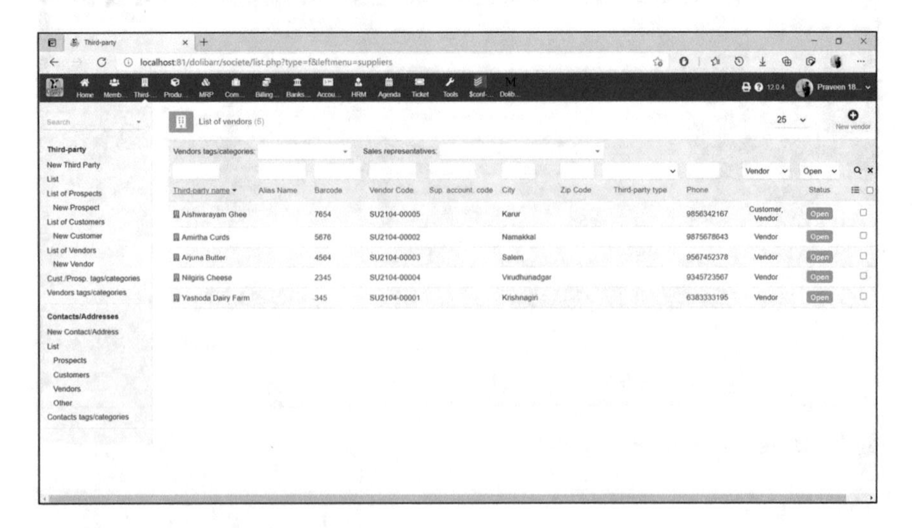

Fig. 10 List of vendors

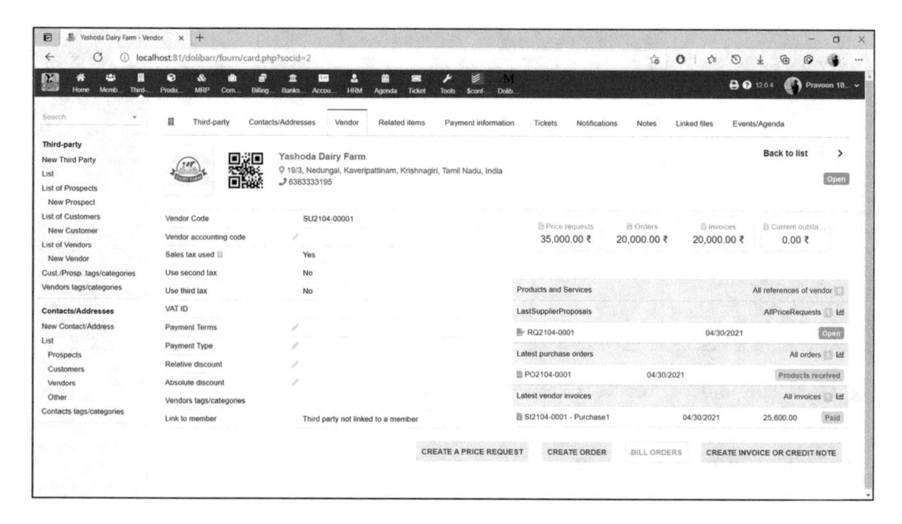

Fig. 11 List of sales order

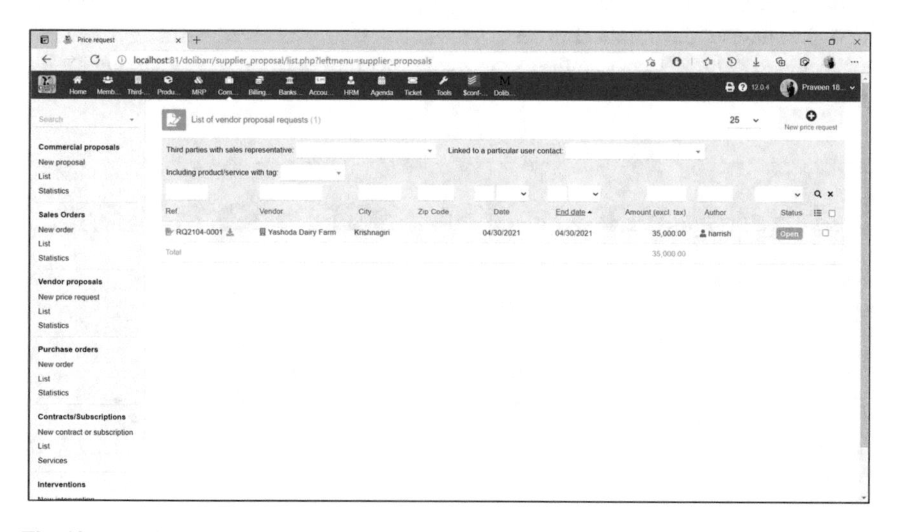

Fig. 12 List of vendors commercial proposal

6.4 Vendor Commercial Proposal

The details of vendor proposals are mentioned here. The details such as Vendor Proposal name, to whom the proposal made, date of the proposal, payment terms are all mentioned. The Product descriptions along with the quantity and price of those products along with the taxes are also mentioned (Fig. 13).

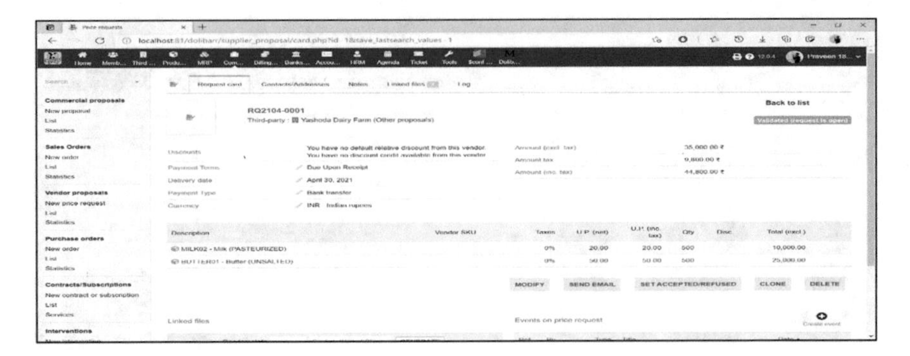

Fig. 13 Details of list of vendors commercial proposal

6.5 *List of Purchase Order*

The list of purchase orders can also be viewed in Fig. 14. Here we can sort out the purchase order with respect to the third parties, order date, planned date, amount. Also, the status of the purchase orders whether it partially received, products received, not received. The Request Author who made this purchase order is also mentioned here.

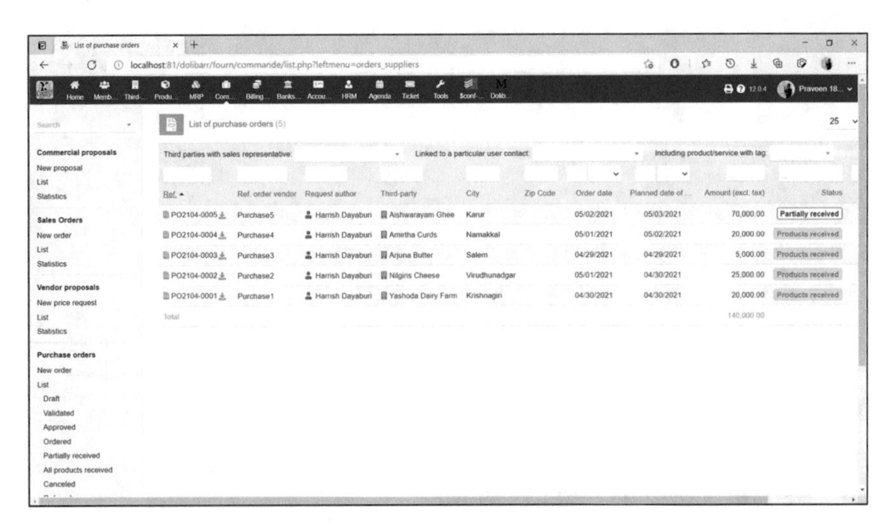

Fig. 14 List of purchase order

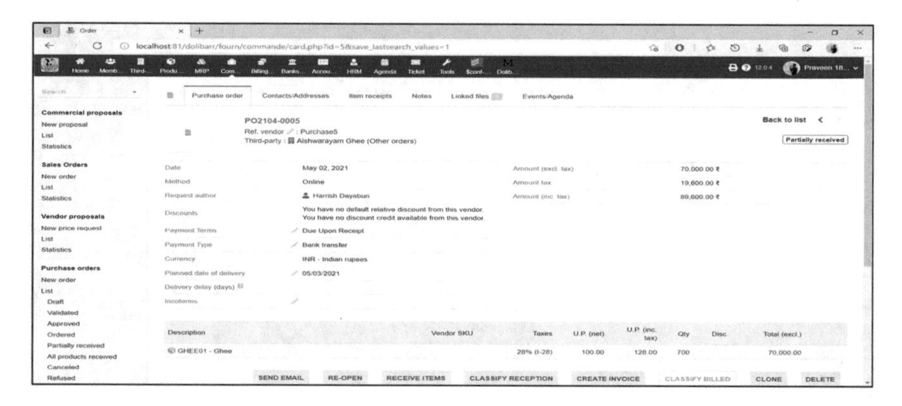

Fig. 15 List of purchase order

6.6 Purchase Order

The details of purchase order are mentioned in Fig. 15. The details such as purchase order id, the provider of the purchase order, date of the ordered, payment terms are all mentioned. The Product descriptions along with the quantity and price of those products along with the taxes are also mentioned.

6.7 Receiving Purchase Order to Different Warehouse

As the purchased order is validated, we need to receive the items/raw materials in the respective warehouse. Here we can either receive all the items in a single warehouse or we can evenly distribute those items to all the warehouse available in Fig. 16. It consists of the products to be received, quantity ordered, quantity dispatched, quantity to dispatch is mentioned.

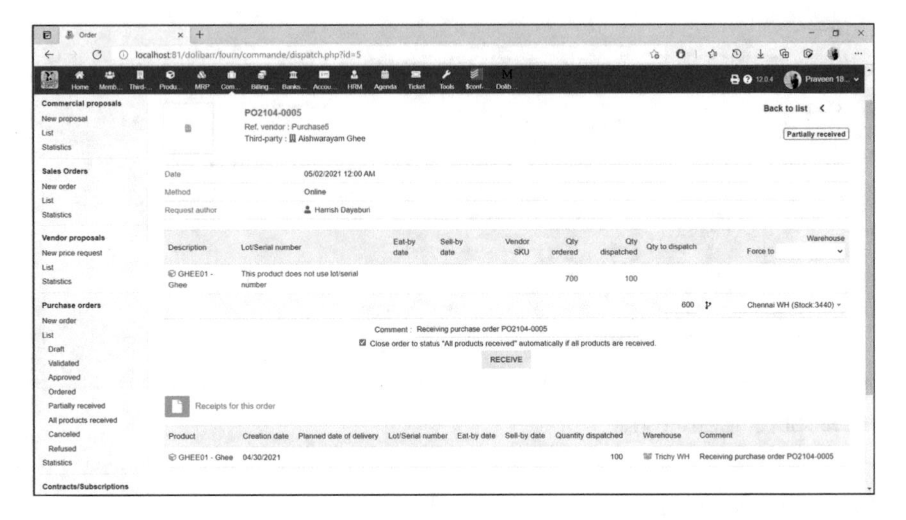

Fig. 16 Details of datawarehouse

7 Financial Modules

7.1 List of Customer Invoices

The list of customer invoices can also be viewed in Fig. 17. Here we can sort out the customer invoices with respect to the Third Parties, Amount, Invoice date, Due date. Also, the status of the customer invoices whether it is Paid, Not Paid, Partially Paid. The Author who made this customer invoice is also mentioned here.

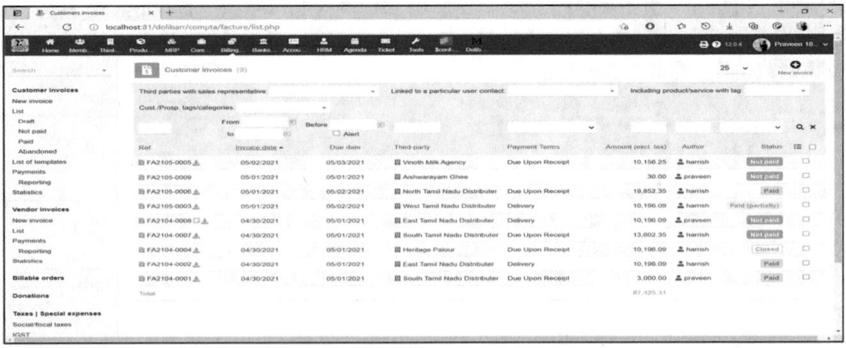

Customer invoices (9)

Ref.	Invoice date	Due date	Third-party	Payment Terms	Amount (excl. tax)	Author	Status
FA2105-0005	05/02/2021	05/03/2021	Vinoth Milk Agency	Due Upon Receipt	10,156.25	harrish	Not paid
FA2105-0009	05/01/2021	05/01/2021	Aishwarayam Ghee		30.00	praveen	Not paid
FA2105-0006	05/01/2021	05/02/2021	North Tamil Nadu Distributer	Due Upon Receipt	19,852.35	harrish	Paid
FA2105-0003	05/01/2021	05/02/2021	West Tamil Nadu Distributer	Delivery	10,196.09	harrish	Paid (partially)
FA2104-0008	04/30/2021	05/01/2021	East Tamil Nadu Distributer	Delivery	10,196.09	praveen	Not paid
FA2104-0007	04/30/2021	05/01/2021	South Tamil Nadu Distributer	Due Upon Receipt	13,602.35	harrish	Not paid
FA2104-0004	04/30/2021	05/01/2021	Heritage Palour	Due Upon Receipt	10,196.09	harrish	Closed
FA2104-0002	04/30/2021	05/01/2021	East Tamil Nadu Distributer	Delivery	10,196.09	harrish	Paid
FA2104-0001	04/30/2021	05/01/2021	South Tamil Nadu Distributer	Due Upon Receipt	3,000.00	praveen	Paid
Total					87,425.31		

Fig. 17 List of customer invoices

7.2 Customer Invoices

The details of customer invoice are mentioned here. The details such as Customer invoice number, for whom the invoice generated, payment terms, payment due date are all mentioned. The Product descriptions along with the quantity and price of those products along with the taxes are also mentioned. The list of vendor invoices can also be viewed. Here we can sort out the vendor invoices with respect to the Third Parties, Amount, Invoice date, Due date. Also, the status of the vendor invoices whether it is Paid, Not Paid, Partially Paid. The Author who made this customer invoice is also mentioned here.

8 Frontal Application

8.1 Poll, Survey or Votes

Polls and votes can be conducted within the ERP system. When a poll is created, it is displayed to all the employees on their respective portals which is shown in Fig. 18. They can cast their poll/votes in their logins. The results can be made public or also as private. We can conduct the poll like "What is the comfortable time to gather for a meeting?".

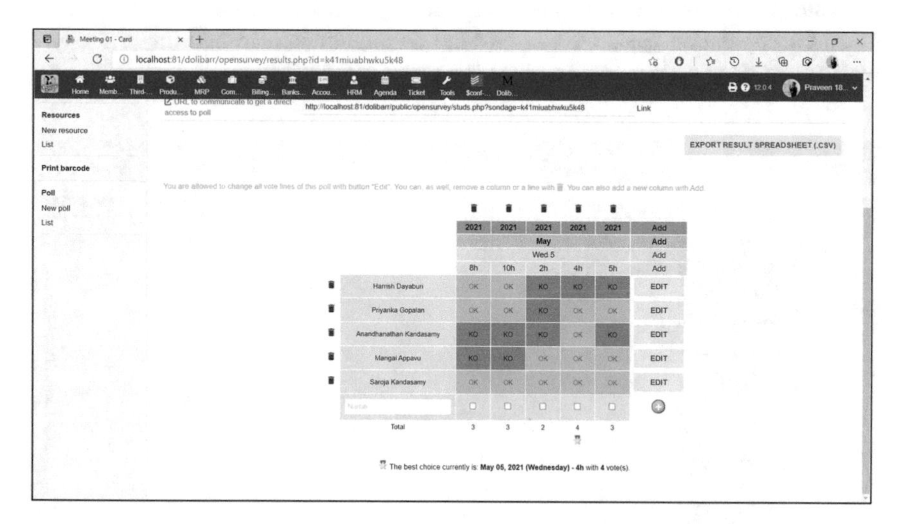

Fig. 18 Details of polls, survey, votes

8.2 E-Commerce Website

Conventionally the sales order should be made by adding a new manual entry directly in the ERP system itself. A simple GUI interface can be added to the ERP system for easy creation of Sales Order which is shown in Fig. 19. TakeOrder provides an interface in the format of forms where we can include the details of the order such as Name, Address, Payment Mode, Name of the Products, Quantity of the products.

We can detect the quality of the products using the Machine Learning Algorithms with Computer Vision. With the help of algorithms such as Convolution Neural Network (CNN) we can extract the information from the image imported and classify them into Good and Spoilt. Here the sample images of dairy products are taken from the internet and they are trained and tested with CNN model that contains a basic net of layers. They provide the results of the images as "Good", "Spoilt". Further we can make use the live webcam feed from the warehouses and can check the quality of the products easily.

8.3 Hotspot Detection for Placing Warehouse

Locating the warehouses are the most important thing should be taken care for a good supply chain which is shown in Fig. 20. All the warehouses should be in the place where most of the industries, customers and people are located. Also, the warehouses we build should not be crowded in a particular region. We need to cover a vast area. Thus, on population basis we can plot the demographic map and find the region where most people are located and we can place the warehouses.

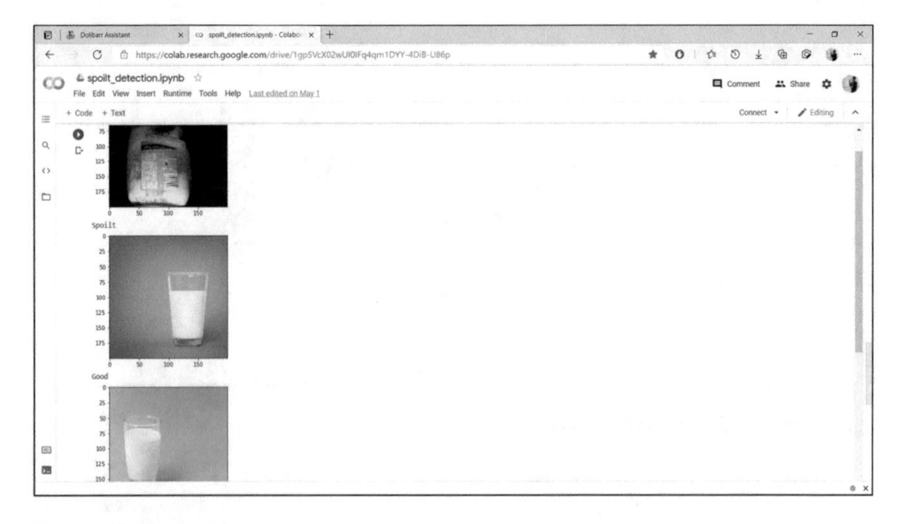

Fig. 19 GUI for ERP

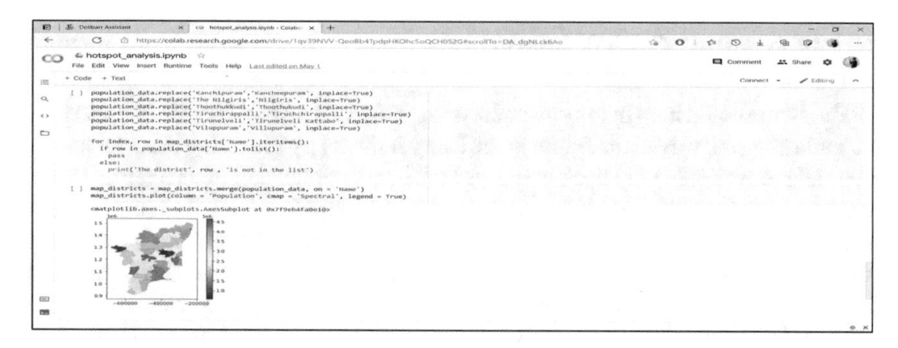

Fig. 20 Placing warehouse

8.4 Chat Bot Assistant

Chatbot assistance powered by Artificial Intelligence helps in understanding the human needs and provide a relevant solution to the need. Dolibarr provides a Chat Bot Assistance that is available in DoliStore. With the help of the Chatbot we can easily create a sales order, purchase order, invoices, add product to the sales order, validate the order, etc. Thus, the chatbot helps in easier creation of orders for those who aren't familiar with the interface and also a time-saving tasks (Fig. 21).

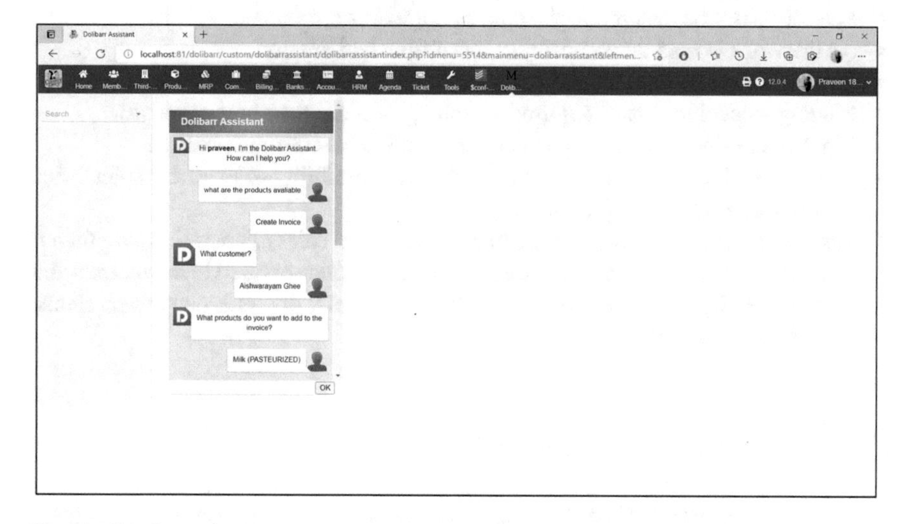

Fig. 21 Chat bot assistant

9 Success and Failure Criteria

ERP system should help in improvement of the DAIRY MANAGEMENT SYSTEM. It should pave the way to the future in the Dairy industry by providing latest technologies and innovations. Way back in the past the conventional dairy management faces challenges such as differences between demand and supply, production supply deadlines, production planning and processing, procurement and distribution, changing demands as customers change their trends, implementation of the accounting and finance, etc. Also, the manufacturers need to maintain the quality standards of the products by frequently check whether the dairy product is spoiled or not. These challenges can't be full-fledgy satisfied by the provided ERP system. Dairy ERP system should be more collaborative with other tools and technologies such IoT devices such as Sensors to monitor the product quality, thus giving a greater advantage. With those additional features along with the basic ERP features such as Procurement, Inventory, Manufacturing, Planning & Scheduling, batch processing, Quality, Accounting, Sales & Distribution, and also provides interactive Dashboards and Reports for all the information and also for the analysis needs [4, 16].

9.1 Problem Statement (Other Than Domain Related)

- Security can't be ensured to the data as the works are done manually.
- As the work is manual it consumes lot of time and energy.
- Records of the customers and vendors are not maintained.
- Finding a specific record required searching of more than half of record.
- Since the system is manual report creation takes lots of time process.
- As the calculations are done manually there is possibility of incorrect calculation resulting in incorrect information.
- Tasks like fee receipt, attendance management and salary generations are manually carried out. In our system, attendance records once entered in maintained and retrieved easily. As we see this reduces the manual work and manpower. Hence processing becomes faster

9.2 Defining Problem Statement

- **Maintaining detail**: There are many more departments in the dairy management such as finance, warehouse, sales, production were all the data maintaining in hand written register is too difficult at time of adding data or retrieving data.
- **Absence of validation**: As the storage and exchange of date is achieved only by user of registers, which lack validation problem.
- **Inquire of available product**: Checking status of the product was performed manually looking in register which Is time consuming.

- **No security facility**: There is no security to handle the departmental data. Anyone can manipulate the data.
- **Update problem**: There is problem with updating or deleting of particular record. For example, they are not able to recollect the details of the previous customers, employees, etc.

9.3 Success of ERP System

- Centralized ERP is provided. In this when one user is updating an information about the products, items, stocks of the dairy products the other user cannot open the same form to avoid conflicts.
- Security to the data is provided by means of Login Form. Only authorized users can have access to the system wherein all users of the system contain their valid user id and password.
- The system allows users to maintain records of customers, suppliers, their orders, raw materials, stock availability as well as bill generation as invoices.
- This system also allows users to generate Customer reports, Supplier reports, sales
- and purchase reports as well as raw material's reports in the form of crystal reports.
- The proposed system is developed to create a simple, user friendly and to use software that avoids the tedious task done by the existing system.
- The proposed system handles the various requirements of the management of the dairy products. The administrator of the system has been assigned to make their settings of the system. He also has the right to see the login details.
- The main requirement of this project is to make the task of inserting the employee related details, product related details and maintain them very simple and time saving.
- After inserting the details, it must be retrieved whenever necessary by search criteria which will give the actual information needed by the valid user.
- There are several reports that are generated based on the employees and room which will show employee information, salary records and attendance. It will also show room check in details and check out details.
- The dairy management system we are designing that helps the dairy product companies to handle those task that had been handled manually.

10 Failure of ERP System

10.1 Standard Price

Every ERP system on creating a product in the system asks for the price of the product. So, that customer can order the dairy product. But in case of dairy product the products produced are easily spoiled. So, most of the products are sold on the debt's way. Even the dairy products vary from price even in a single day. Milk is higher the price when it was sold in afternoon rather than morning. So, a standard price can't be fixed for Dairy products.

10.2 Inaccurate Costing

Dairy products include not only the manufacturing of the dairy product. It also includes the covers, packet, bottles, sealing. If we mention all those as a product in manufacturing a dairy product the cost of the product will be more. Thus, we can't add up those residual prices while calculating the overall price which leads to inaccurate costing of the products. Later those items like covers, packets should be included separately and should be calculated for a batches of manufacturing products.

10.3 Avoiding Raw Materials Prices

Some of the raw materials such as Rennet and Cultures in making cheese is also to be considered while fixing the price of the cheese. But those raw materials are not for sales. Thus, we can't fix the selling price of those raw materials. Then it leads in avoiding those prices when fixing the Cheese price.

11 Conclusion

Through the implementation of project, we found that Conventional Dairy Management is not enough to include all the facilitates provided for dairy management in main stream. It is found that we need an ICT (Information communication technology) based ERP system to integrate geographically all the warehouse and distributers of the Dairy products. Using different technologies like GPS, RFID, IOT, remote sensing, android mobiles ERP can gather and store data in central database. Through central database at different level user can generate different reports. Through different reports workers of the dairy management system can improve its decision making. These reports help cooperative to improve overall sales orders, productivity etc. On worker's side ERP should have modules like milk management, payment tracking, weather alert, expense tracking, market place and important contacts. On cooperative's side ERP should have modules like raw materials collection management, inventory management, milk transit management, sales, human resource management, customer relationship management, and accounting.

Thus, I can conclude that ICT based ERP modules can significantly improves efficiency, effectiveness and integrate them in main stream. ERP will also improve overall production, transparence and efficiency of dairy product workers. ICT based ERP modules will become a stepping stone for the revival for whole Dairy Management System.

References

1. Talukder, B., Agnusdei, G.P., Hipel, K.W., Dubé, L.: Multi-indicator supply chain management framework for food convergent innovation in the dairy business. Sustain. Futures **3**, 100045 (2021)
2. Manikas, I., Manos, B.: Design of an integrated supply chain model for supporting traceability of dairy products. Int. J. Dairy Technol. **62**(1), 126–138 (2009)
3. Augustin, M.A., Udabage, P., Juliano, P., Clarke, P.T.: Towards a more sustainable dairy industry: integration across the farm–factory interface and the dairy factory of the future. Int. Dairy J. **31**(1), 2–11 (2013)
4. Van Berkum, S.: Integrating into the EU: challenges for the Serbian dairy supply chain (No. 1338–2016–103973) (2010)
5. van Duren, E., Sparling, D.: Supply chain management and the canadian agri-food sector. Can. J. Agric. Econ. Revue Canadienne d'agroeconomie **46**(4), 479–489 (1998)
6. Fatorachian, H., Kazemi, H.: Impact of industry 4.0 on supply chain performance. Prod. Plan. Control **32**(1), 63–81 (2021)
7. Samaranayake, P., Laosirihongthong, T.: Integrated model of supply chain processes and performance measures: a case of dairy industry. In: Proceedings of the 28th Australian and New Zealand Academy of Management Conference (ANZAM 2014): Reshaping Management for Impact, 3–5 December 2014, UTS Business School, Sydney, Australia (2014)
8. Ferreira, F.U., Robra, S., Ribeiro, P.C.C., Gomes, C.F.S., Almeida, J.A.D., Rodrigues, L.B.: Towards a contribution to sustainable management of a dairy supply chain. Production 30 (2020)
9. Emidio, J., Lima, R., Leal, C., Madrona, G.: How can mixed integer linear programming assist dairy manufacturers by integrating supply decisions and production planning? J. Agribus. Dev. Emerg. Econ. (2021)
10. Susanty, A., Bakhtiar, A., Jie, F., Muthi, M.: The empirical model of trust, loyalty, and business performance of the dairy milk supply chain: a comparative study. Br. Food J. (2017)
11. Ruben, R., Dekeba Bekele, A., Megersa Lenjiso, B.: Quality upgrading in Ethiopian dairy value chains: dovetailing upstream and downstream perspectives. Rev. Soc. Econ. **75**(3), 296–317 (2017)
12. Faisal, M.N., Talib, F.: Implementing traceability in Indian food-supply chains: an interpretive structural modeling approach. J. Foodserv. Bus. Res. **19**(2), 171–196 (2016)
13. Tian, F.: A quality and safety control system for China's dairy supply chain based on HACCP & GS1. In: 2016 13th International Conference on Service Systems and Service Management (ICSSSM), pp. 1–6. IEEE (2016)
14. Jachimczyk, B., Tkaczyk, R., Piotrowski, T., Johansson, S., Kulesza, W.: IoT-based dairy supply chain-an ontological approach. Elektronika ir Elektrotechnika **27**(1), 71–83 (2021)
15. Bhardwaj, A., Mor, R.S., Singh, S., Dev, M.: An investigation into the dynamics of supply chain practices in Dairy industry: a pilot study. In: Proceedings of the 2016 International Conference on Industrial Engineering and Operations Management, Detroit, Michigan, USA, pp. 1360–1365 (2016)
16. Sel, Ç., Bilgen, B.: Quantitative models for supply chain management within dairy industry: a review and discussion. Eur. J. Ind. Eng. **9**(5), 561–594 (2015)

Data Protection Using Multiple Servers for Medical Supply Chain System

Nikhilesh Saragadam, Suri Koushmitha, Yerru Nanda Krishna Arun, and Chiranji Lal Chowdhary

Abstract Nowadays most of the organizations are experiencing data breaches either from the insiders or externally from outside of the organization. Therefore, at the exact time patient's personal health data would not be circulated here and there. As data leakage can lead to huge loss of money and create lot of issues in the organization. Even though they are implementing certain security mechanisms but they fail to store them in a secured manner. Present inventions can only be able to protect the data during the transmission stage, but cannot stop the insider attacks. Sometimes they are also chances; here the manager of the patient database can reveal the sensitive information. Therefore, in order to overcome all this issues, we propose a model to get rid of inside attacks with the help of multiple servers to store patient's data. Our main contribution is to securely distribute the patient data in numerous data servers which employed the Paillier and ElGamal cryptosystems for performing statistical study on the patient data without cooperating the patients' confidentiality.

Keywords Crypto systems · Statistical analysis · Transmission · Security mechanisms

1 Introduction

Electronic health record is treated as an electronic version of the medical record of the concerned person who is visiting to the hospital. These records will be maintained and managed by the health sector. Now a days we are arriving into a situation where on one side the advancements in the domain of health are rapidly increasing day by day and on a parallel world we are observing the medical data getting leaked due to the drawbacks that were found in the medical technologies [1].

Existing solutions can able to protect patient during the level of transmission, but cannot able to handle if there are any insider attacks taking place. So, after noticing the loopholes we would like to propose a solution to shield the medical

N. Saragadam · S. Koushmitha · Y. N. K. Arun · C. L. Chowdhary (✉)
School of Information Technology and Engineering, Vellore Institute of Technology, Vellore, India

K. Perumal et al. (eds.), *Innovative Supply Chain Management via Digitalization and Artificial Intelligence*, Studies in Systems, Decision and Control 424,
https://doi.org/10.1007/978-981-19-0240-6_11

information which gives the assurance to patient as well as doctor. We deployed a practical approach in order to prevent inside attack with the usage of required number of servers to upload and retrieve the data. Making use of two encryptions [2] helped us to add an additional layer of security to our proposed system. In cryptography terminology we term this double encryption as hybrid encryption [3]. This technique is mostly made use, especially now-a-days any organization or sector definitely needs security of their data.

Irrespective of domains, we need security to access or store the data in the database or cloud based upon the requirement. So, we need proper monitoring whether the data gets leaked or stolen due to involvement of third parties or inside attacks. Doctor' page will be having the options to retrieve all the details from both sensor and server. After the retrieval process is done by the doctor, he needs to perform two descriptions [4], so that only he can access the prescription of the concerned patient. Inclusion of digital signature created a one more layer of security to our proposed system. Principle behind the inclusion of this in our system is, doctor need to have the assurance that the data he is retrieving is from a trusted source, so makes use of this signature [5] to validate and verify it to provide the assurance of the data privacy. This can be alphanumerical.

The main goal of our framework is to provide secure patient-centric access and efficient key management at the same time. So, our moto is improving healthcare quality by creating a more trust between consumers and their healthcare providers and third-party organizations. This system acts like a solution in two-way mechanism, which can help the organization either from the inside attacks [6] or also the attacks that are taking place from the outside sources which we usually call them as hackers. We are in a situation where the advancements are getting changed for every hour within the country, so we badly need a system to secure as much as possible in order to avoid the attacks that may create a huge loss to any of the organization.

The main contributions of this chapter are:

- Retrieval of information with the help of Hybrid encryption using Elgamal and Paillier algorithms.
- Paillier is found more beneficial while performing the decryption due to its nature of additive homomorphism.
- Proposed system can have the ability to act as a barrier to keep a hold on insider attacks due to the double encryption techniques imposed in our system.
- The results depict that this proposed system is meant for controlling the data leakage from both inside and outside threats.

The organization of the chapter has the sequence of information as follows—Sect. 2 of the chapter provides an explicit review of related works and Sect. 3 describes the preliminaries, along with the basic discussion of the techniques and algorithms used in this work. Section 4 discusses the result analysis and Sect. 5 collates the conclusion and highlights the direction of future.

2 Related Work

Lim et al. promulgated (an idea or cause) an article on security issues on Wireless Body Area Network for Remote Healthcare Monitoring [1]. Core components that were included by the authors in the system are inclusion of biosensor pairs, hardware elements to automate the setup process of wireless body area network, sending data to an internet server, automation of things like data collection, profiling and then alerting to report it. This system allows monitoring the health status of patient's, upon which decision will be framed to make the treatment. Decisions will be made with the help of sensors. They included a crucial part in their system which will help the patients to view their nearby hospitals, availability of doctors in their particular region. In rider to deploy security features in their system, they made use of Elliptic curve cryptographic algorithm and protocols [7] like mutual authentication and group key agreement.

Elgamal proposed a new Public Key Cryptosystem and a Signature Scheme based on the difficulty of computing discrete logarithms over finite fields. They proposed a signature-based technique with parallel execution of Diffie-Hellman [2] key distribution whose goal is to reach the public key cryptosystem. They found that now-a-days, most of the systems are depending on difficulty of computing logarithms over certain finite fields. Initially they will be performing randomization in enciphering operation [8], here the obtained cipher text for a message is not at all repeated. This suggests that if we are trying to encipher the same message two times, then we cannot expect the outcome to be same cipher text. They said that this statement strongly assures that it can prevent attacks like probable text attack. As enciphering operation is playing a crucial role, it needs two exponentiations, whereas for deciphering operation need only one exponentiation.

Sang et al. [4] conducted a survey on Data aggregation in wireless sensor networks. There were certain issues that are addresses during the deployment of sensor network in a hostile environment. They are data aggregation [4], security issues and data confidentiality [9]. Out of all these, data aggregation is most widely used in wireless sensor networks. So, taking this into consideration they have divided into two cases which are hop-by-hop encrypted data aggregation and end-to-end encrypted aggregation. Benefits they found with end-to-end encrypted data aggregation are it can address the vulnerability queries; it can aggregate the encrypted sensor readings without the involvement of decryption. Based on the results obtained, they came a conclusion that end-to-end encrypted data aggregation has more computation cost with respect to sensor nodes, but it is achieving more security when compared with hop-by-hop encrypted data aggregation.

Han et al. proposed a privacy-preserving and multifunctional health data aggregation (PPM-HDA) mechanism with fault tolerance for cloud-assisted WBANs. The proposed technique will help the cloud servers to compute multiple statistical functions of certain user's health data in a secure way in order to offer various services. In the initial stage, they proposed multifunctional health data additive aggregation

(MHDA) technique [5]. Reason behind this is they said it can support some additive aggregate functions such as average and variance. In order to cover the other non-additive aggregations such as min, max, median, percentile and histogram, they found the extension for the technique which is MHDA+. Proposed system can able to withstand many attacks, which most of them are suffering from data aggregation techniques. From the obtained results they found that the computational overhead of MHDA+ is decline with the assistance of cloud servers and also found that MHDA+ is more efficient than the existing aggregation techniques with respect to communication overhead when the application is in need with large plaintext space.

Martí et al. presented the security Services for Mobile e-Health Services and how they have implemented them. They proposed certain security services that are needed for mobile healthcare applications. These services need to overcome the threats related to confidentiality, integrity, authentication, non-repudiation and access control. Security mechanisms [6] can be included in different communication layers, which help in providing different security features depending on the layer where security is getting involved with. Layers can be data link layer, network layer, transport layer and application layer. The benefits the authors observed while imposing these security mechanisms are use of standard user-oriented security techniques, no use of IPsec host security and all communications from terminal to host are secured with the involvement of authentication and encryption.

He and Zeadally made an analysis on RFID authentication schemes for Internet of Things (IOT) in healthcare environment using elliptic curve cryptography. They discussed about the security requirements that were required for Radio Frequency Identification (RFID) [10] authentication schemes with respect to performance and security. RFID is treated as one of the core technologies in the healthcare sector. Present days, Elliptic Curve Cryptography (ECC)-based RFID authentication schemes are more involved due to its high security. If any healthcare organization is in need with IOT implementation then they need RFID authentication which is treated as a critical security service. In order to ensure secure communication in any RFID system, it is mandatory to construct suitable security model for ECC-based RFI techniques. After this step, we need to design ECC-based RFID authentication techniques which can be probably help to add an additional layer of security to the model which we designed.

He et al. proposed a lightweight system to secure wireless medical sensor networks. They proposed a lightweight and secure system for Wireless Medical Sensor Networks [11]. These are not mostly used in e-healthcare which allows the data of a particular patient' body parameters will be collected with the help of implantable biosensors. Bi, now a days we can observed the data which we are collecting is not at all secure, so they proposed this secure system. It makes use of hash-chain based key updating technique and proxy-protected signature method in order to achieve effective secure transmission and proper data-access control. They also added two more features which are to provide backward secrecy and privacy preservation. If there is a use of low-power sensor nodes then suitable and appropriate symmetric-key encryption, decryption [12] and hash operations need to be impose on the system.

Anees Ara. Jr, Mznah rodhaan, Yuan tian and Abdullah proposed a secure privacy-preserving data aggregation scheme based Bilinear ElGamal Cryptosystem for Remote Health Monitoring Systems [13]. They proposed a Secure Privacy-Preserving Data Aggregation scheme based on bilinear pairing for any remote health monitoring systems in order to improve the data aggregation efficiency and data privacy. It makes use of homomorphic property of the Bilinear Elgamal cryptosystems in order to perform secure computation and includes it with the aggregate signature methods, thereby enabling data authenticity and data integrity in wireless body area network. This proposed system is proved that it is semantically secure under the Decisional Bilinear Diffie-Hellman (DBDH) assumption considered. Security Analysis made on this system depicted that this system can preserve data confidentiality, data authenticity [14] and data privacy and also resist some attacks like eavesdropping and replay attacks.

Thilakanathan et al. [15] presented the attempts to address the issues of privacy and security in mobile telecare and cloud computing [16] domains. They deployed a telecare application which will allow the doctors to monitor the patients in a virtual mode. This acts as a medium between the patients and doctors to share the information. Key features that were related to this application are ability to handle large datasets and efficient user revocation. They made an analysis on the security aspects of the protocol designed and performed some tests to test the feasibility. Usage of cloud [15] in this application is due to its combination with mobile technologies has enabled the doctors to monitor the patients conveniently and assess their records whenever possible.

3 Preliminaries

Algorithms considered to perform encryption and decryption are Elgamal and Paillier. As we ae in need of high security we have implemented this hybrid encryption concept.

3.1 Elgamal Algorithm

In most of the chapters referred, the authors suggested to make use of this algorithm especially for hybrid encryption. It depends on certain parameters which can affect the performance, speed and security of this algorithm. So, the value of parameters that we consider should be made properly to deduct the complexity and make the system in a secure manner. In this mechanism, a single plaintext can be encrypted to many numbers of possible ciphertexts. So, to be represented in a radio format it will be 2:1 from plaintext to ciphertext.

Here, the starting with the generation of keys [17] which will be randomly done by the system in the back-end. Selected files or data that need to store somewhere

Table 1 Parameter description for Elgamal algorithm

Symbol	Definition
r	Large prime number
R	First part of decryption key
A1	Second part of decryption key
A2	Third part of encryption key
I	Random integer
B1	First part of cipher text
B2	Second part of cipher text

need to be encrypted. After the encryption process is done, it's up to the user to send it to other systems or to upload it in the servers. At last, in order to decrypt he need a private key to view the file.

Algorithm. This algorithm majorly contributes to three parts. They are
- Key Generation
- Encryption
- Decryption

(a) Key Generation.
i. Select large random prime number (r)
ii. Select first part of decryption key or private Key(R)
iii. Select second part of encryption key or public key(A1)
iv. Third part of the encryption key or public key (A2).
A2 = A1R mod r
v. Public key = (A1, A2, P); Private key = R
(b) Encryption.
i. Select Random Integer I
ii. B1 = A1R mod r
iii. B2 = (Plain Text * A2 R) mod r
iv. Cipher Text = (B1, B2)
(c) Decryption.
Plain Text = [B2 * (B1R) − 1]mod r
(d) Parameter description in Table 1.

3.2 *Paillier Algorithm*

The Paillier algorithm monitors a development method called key pair-based cryptography. This means that user gets a public key and also a private key, and messages encrypted with their respective public key were able to decrypt with their private key. This algorithm makes use of additive homomorphism [13] which means the use of splitting module arrives here. Suppose, if the user is sending the file which is divided

Table 2 Parameter description for Paillier algorithm

Symbol	Definition
a, b	Prime numbers
n, p	Acts as public key
λ, μ	Acts as private key
c	Cipher text
y	Message
m	Paint text

into 'n' number of files, then while perform to decrypt, such messages are additional to be correct.

Starting with the generation of keys which will be randomly done by the system in the back-end. Selected files or data that need to stored somewhere need to be encrypted. After the encryption process is done, it's up to the user to send it to other systems or to upload it in the servers. At last, in order to decrypt he need a private key to view the file.

Algorithm. This algorithm majorly contributes to three parts. They are
- Key Generation
- Encryption
- Decryption

(a) Key Generation.
i. Choose two prime numbers a & b and calculate $n = a * b$ and
$\lambda = \text{LCM}(a - 1, b - 1)$ such that $\text{GCD}(a * b, (a - 1) * (b - 1)) = 1$
ii. Select p Є Zn2 and calculate $\mu = (L(p\ \lambda \bmod n\ 2)) - 1 \bmod n$ where $L(x) = x - 1/n$
iii. n, p acts as a public key or encryption key
iv. λ, μ acts as a private key or decryption key
(b) Encryption.
i. Let y Є Zn be the message
ii. Choose x Є Z * n
iii. Required Cipher text is
$c = g * y * x\ n \bmod n\ 2$
(c) Decryption.
Compute $m = L(c\ \lambda \bmod n\ 2) * \mu \bmod n$
(d) Parameter description in Table 2.

3.3 Proposed Architecture

This section is having a detailed discussion on the architecture of the proposed system.

- Initially doctor is going to perform Elgamal key generation which is with respect to generation of public key and private key.
- Now, we are going to execute the server page, here user can consider 'n' number of servers based upon our requirement. User need to enter the server id, so that it makes sense about the unique servers considered so far. This Server will be connected with the help of public key (Elgamal) generated by the doctor.
- In this distributed server model, first the server nodes and sensor nodes will be connected with the doctor.
- Now entering into the sensor node, here sever details will be displayed and file will be encrypted and gets uploaded. This connection for the sensor node will be done using Paillier public key. Only after performing encryption the files get uploaded, if not they don't get uploaded and instructs to encrypt it first.
- The task at the end should be again made by the doctor who is to view the particular patient's prescription which can be done only when he performs Elgamal and Paillier decryptions. Doctor can also able to access the details of servers and senor nodes.
- Doctor's job is to first download all the data from the respective servers. Now, first decrypts the data based on Elgamal decryption with the help of Elgamal private key. After this, he should decrypt the data based on Paillier decryption using Paillier private key. Though the data is divided in the servers but this will not affect the data received at doctor's end due to the factor additive homomorphism (Fig. 1).

Tabular Data. When a patient is visiting a hospital, he/she needs to undertake the preliminary test and then the results need to copy to the prescription. This prescription will be having the data of he concerned patients. This data can be of any type based upon the wish of the receptionist to enter the data into the prescription in Table 3.

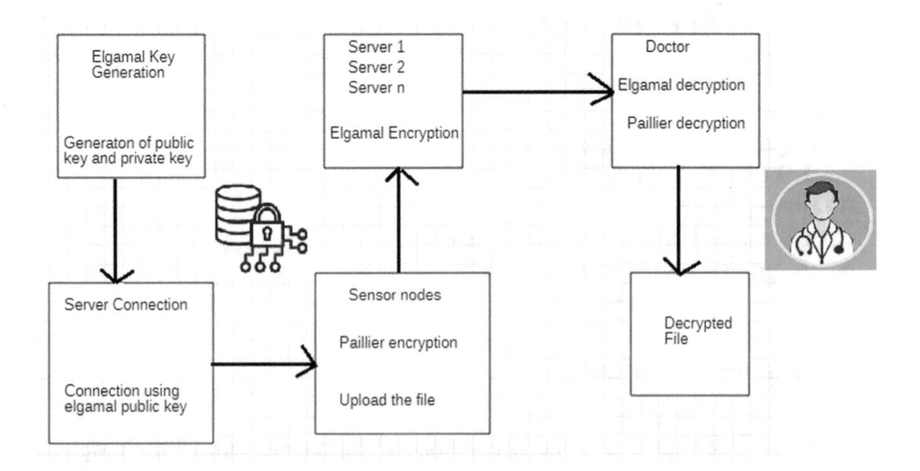

Fig. 1 Architecture of supply chain based medical system

Table 3 Protection of information and exchange of asset with respect to type of data

Type of data	Privacy protection	Asset exchange
Manual data	Low	Low
Scanned data	Medium	High
Electronic data	High	High

Table 4 Attack types and its protection level

Attack type	CIA triad	Protection
Hacking	Confidentiality, integrity	High
Social engineering	Confidentiality	Low
Data theft	Confidentiality	High
Man in the middle attack	Confidentiality, integrity, availability	Medium

Protection will be more to the electronic form of data due to the clear picture of data in this format respectively.

Even though we are creating a secured portal with the security mechanism imposed on the system, we can observe some malicious intruders who try to steal our privacy data in Table 4. Type of attacks that can be encountered and its appropriate protection level is mentioned here in the below table.

4 Experimental Setup, Result Analysis and Discussion

4.1 Experimental Setup

Figure 2 describes the experimental setup of proposed work.

- Entering into the practical implementation of the project, Doctor's page will be consisting of key generation (public and private), sensor, server and file details.
- In Server's page we need to establish a connection with the doctor, so that only he can able to access the sensor details and upload the files based on the requirement to store for security.
- Sensor's page includes connection establishment, viewing server details and to encrypt and upload the file.

4.2 Results

Figures 3, 4 and 5 are showing the results. Connection need to be made with the public key of doctor which needs to be matched with the public key in the server

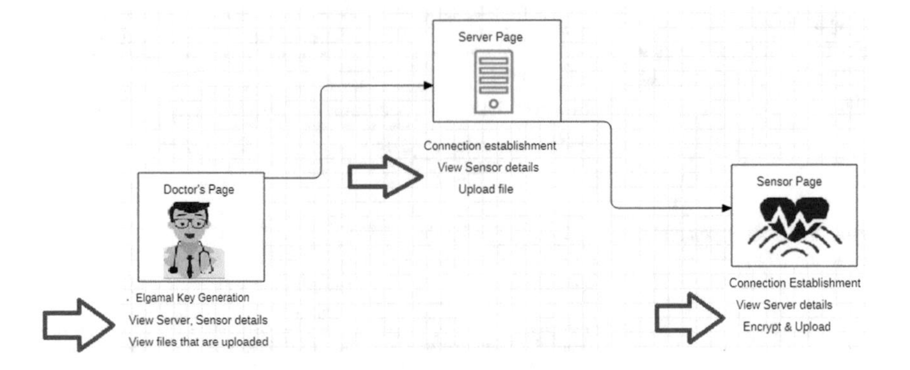

Fig. 2 Experimental setup of supply chain medical system

Doctor's Page

Generation of Elgamal Key | Retrieve Server Info | Retrieve Sensor Info | View files

Elgamal Public Key:

Elgamal Private Key:

Generate Key's

Fig. 3 Doctor's page

Server's Page

Connection phase | Retrieve Sensor Node Info | File Info

Elgamal Public Key:

Connection Establishment

Fig. 4 Server's page

node. Here we can view the sensor and file details. Connections need to be established and then we need to upload the file in the server.

Encryption needs to be performed and then only it can be uploaded in the server. Encryption to be followed is paillier. We need to retrieve files from sensor and then

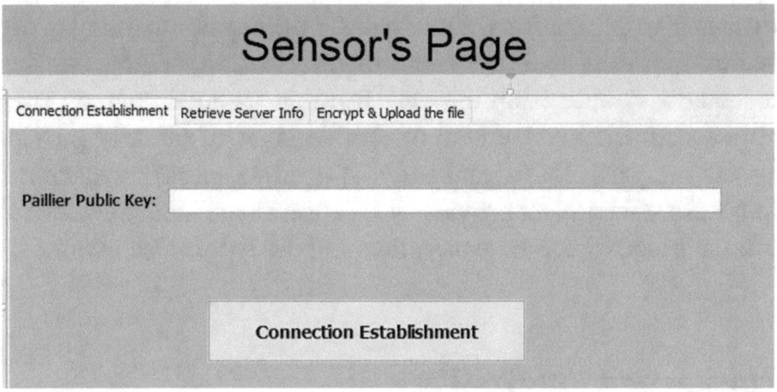

Fig. 5 Sensor page

another encryption (elgamal) need to be performed, thereby adding an additional layer of security to this particular system. Now if doctor is need to download the file, and then we need to download all the files from server. We need to perform two decryptions in order to download the encrypted file.

4.3 Performance Analysis

Encryption we made use of which are Elgamal and Paillier played a crucial role in the implementation phase. Reason behind performing this hybrid encryption is to add an additional layer of security [18] to the system. In the sensor node, while uploading the file into the server it performs Paillier encryption. At the server node, it will be first checking the files that got received and then perform the Elgamal encryption before sending it to the doctor. So, our plan of making use of two algorithms added more reliability, so while coming to comparison analysis, we can strongly say that both are equally performed to achieve the goal.

4.4 Applications

Health Sector. Since, we can observe the scams that were been encountered in our society in this sector are getting huge day by day. Patient need to be aware about his data which is in prescription format, such that it should reach to the doctor's end without any data leakage in between the path the prescription travelled. So, we try to implement a system to all the users who need to be aware by protecting their health data. Until the data enters into the doctor's hand nobody can hack or stole the data, even if they try to do, they can't able to decode the information of the respective patient.

Medium sized organizations. Employees working in a particular organization may face certain barriers like disclosure of information, selling our data to third parties and make a deal to hack the data from an external source. These hurdles can be stopped with the involvement of double layer of security providing to the files/data/assets the particular organization is need to access or make use of it for future purpose. So, this project provides a solution to maintain the data in a secured manner with the usage of certain encryption and decryption techniques.

5 Conclusion and Future Work

So, in order to conclude we have developed a system where it accepts the details of the patient and through the server and sensor nodes the encryption process has been done and all the files are redirected to doctor's page and there by decryption process has been tested out and files have been downloaded successfully. For the real client (e.g., clinical scientist) to perform factual examination on the patient information, proposed some new conventions for normal, connection, difference and relapse investigation, where the three information servers co-work to process the patient information without unveiling the patient protection and afterward give the client the measurable examination results. Security and protection investigation has indicated that conventions are secure against both outside and inside.

References

1. Lim, S., Oh, T.H., Choi, Y.B., Lakshman, T.: Security issues on wireless body area network for remote healthcare monitoring. In: 2010 IEEE International Conference on Sensor Networks, Ubiquitous, and Trustworthy Computing, pp. 327–332. IEEE (2010)
2. ElGamal, T.: A public key cryptosystem and a signature scheme based on discrete logarithms. IEEE Trans. Inf. Theory **31**(4), 469–472 (1985)
3. Chowdhary, C.L., Patel, P.V., Kathrotia, K.J., Attique, M., Perumal, K., Ijaz, M.F.: Analytical study of hybrid techniques for image encryption and decryption. Sensors **20**(18), 5162 (2020)
4. Sang, Y., Shen, H., Inoguchi, Y., Tan, Y., Xiong, N.: Secure data aggregation in wireless sensor networks: a survey. In: 2006 Seventh International Conference on Parallel and Distributed Computing, Applications and Technologies (PDCAT'06), pp. 315–320. IEEE (2006)
5. Han, S., Zhao, S., Li, Q., Ju, C.H., Zhou, W.: PPM-HDA: privacy-preserving and multifunctional health data aggregation with fault tolerance. IEEE Trans. Inf. Forensics Secur. **11**(9), 1940–1955 (2015)
6. Gomathy, C.K., Geetha, V., Bhargavi, M., Jayanthi, T.: A medical information security using cryptosystem for wireless sensor networks
7. Azaria, A., Ekblaw, A., Vieira, T., Lippman, A.: Medrec: Using blockchain for medical data access and permission management. In: 2016 2nd International Conference on Open and Big Data (OBD), pp. 25–30. IEEE (2016)
8. Elhoseny, M., Ramírez-González, G., Abu-Elnasr, O.M., Shawkat, S.A., Arunkumar, N., Farouk, A.: Secure medical data transmission model for IoT-based healthcare systems. IEEE Access **6**, 20596–20608 (2018)

9. Jin, H., Luo, Y., Li, P., Mathew, J.: A review of secure and privacy-preserving medical data sharing. IEEE Access **7**, 61656–61669 (2019)
10. He, D., Zeadally, S.: An analysis of RFID authentication schemes for internet of things in healthcare environment using elliptic curve cryptography. IEEE Internet Things J. **2**(1), 72–83 (2014)
11. He, D., Chan, S., Tang, S.: A novel and lightweight system to secure wireless medical sensor networks. IEEE J. Biomed. Health Inform. **18**(1), 316–326 (2013)
12. Chen, F., Luo, Y., Zhang, J., Zhu, J., Zhang, Z., Zhao, C., Wang, T.: An infrastructure framework for privacy protection of community medical internet of things. World Wide Web **21**(1), 33–57 (2018)
13. Ara, A., Al-Rodhaan, M., Tian, Y., Al-Dhelaan, A.: A secure privacy-preserving data aggregation scheme based on bilinear ElGamal cryptosystem for remote health monitoring systems. IEEE Access **5**, 12601–12617 (2017)
14. Puppala, M., He, T., Yu, X., Chen, S., Ogunti, R., Wong, S.T.: Data security and privacy management in healthcare applications and clinical data warehouse environment. In: 2016 IEEE-EMBS International Conference on Biomedical and Health Informatics (BHI), pp. 5–8. IEEE (2016)
15. Thilakanathan, D., Chen, S., Nepal, S., Calvo, R., Alem, L.: A platform for secure monitoring and sharing of generic health data in the Cloud. Future Gener. Comput. Syst. **35**, 102–113 (2014)
16. Sun, W., Cai, Z., Li, Y., Liu, F., Fang, S., Wang, G.: Security and privacy in the medical internet of things: a review. Secur. Commun. Netw. (2018)
17. Yi, X., Bouguettaya, A., Georgakopoulos, D., Song, A., Willemson, J.: Privacy protection for wireless medical sensor data. IEEE Trans. Dependable Secure Comput. **13**(3), 369–380 (2015)
18. Qiu, H., Qiu, M., Liu, M., Memmi, G.: Secure health data sharing for medical cyber-physical systems for the healthcare 4.0. IEEE J. Biomed. Health Inform. **24**(9), 2499–2505 (2020)

Printed in the United States
by Baker & Taylor Publisher Services